Indi

Industrial Designer
The Artist as Engineer

by W. Dorwin Teague

Armstrong World Industries, Inc.
Lancaster, Pennsylvania

copyright © 1998 by W. Dorwin Teague

no part of this book may be reproduced or
transmitted in any form or by any means,
electronic or mechanical, including photocopy,
recording or any information storage and retrieval
system now known or to be invented, without
permission in writing from the author
except by a reviewer who wishes to quote brief
passages in a review for inclusion in a
magazine, newspaper or broadcast.

book produced by Beverly Rae Kimes

ISBN 0-9667313-C-1

Armstrong World Industries, Inc.
313 West Liberty Street
Lancaster, PA 17603

for
Harriette

Prologue

This book is both a history of industrial design and an autobiography by one who has been in the profession since its inception in the late twenties. My timing was right. I was born in 1910 and by 1929 I was nineteen, ready to design a car body which was received enthusiastically at the time and is regarded as a classic today.

Designers who specialize in one category, such as furniture, pottery, dresses, graphics or houses have been busy since prehistoric days. Some of the furniture buried with Egyptian pharaohs in 3000 B.C. is at least equal in form and function to anything produced today. Museum visitors are amazed at the beauty of the weapons made by prehistoric flint knappers. But when industry began to produce office equipment, telephones and domestic appliances faster and less expensively, manufacturing capabilities outstripped the ability to forecast what the market wanted. Guidance was needed to improve the aesthetics and ergonomics of products, to make them more competitive. There was no established guild to which the industry could go for help, except for the new industrial designers. One or two early experiences with these designers, such as Eastman's instant success with my father's Baby Brownie camera in the early thirties, and industry became a believer.

The fact that America was entering the Great Depression was no obstacle to the new profession. If anything, it actually helped industrial designers. The large corporations with enough capital to weather the storm were looking for any measure that could pilot them to a safe and prosperous future. The top designers were astute enough to demand substantial fees; if a company objected, no matter, there were plenty of others who would gladly accept. Banding together to establish a group called the Society of Industrial Designers (my father was the first president), they realized that a certain degree of integrity was essential to continuing prosperity. By-laws were passed against plagiarism and piracy of clients. The designers sincerely tried, and usually succeeded, in giving clients their money's worth.

The main difference between the new industrial designers and earlier groups (furniture designers, clothing designers, architects, et al.) was that this new organization was faced with the mass production of products for which no responsible guild existed. None of the existing designer categories quite filled the bill. Probably architects, who were aware of the importance of aesthetics and ergonomics, came closest. They could, and did, design furniture, lighting and fixtures but were generally too busy with the drawings, specifications and construction of their houses to worry about details of toasters, furnaces and laundry equipment.

Of all the early industrial designers, except for Raymond Loewy and this writer, none had much or any engineering training. Henry Dreyfuss was a stage designer, as was Norman bel Geddes. Donald Deskey was a painter and furniture designer. Russell Wright also trained as a painter, and my father was a commercial artist, working in advertising and graphics. This very lack of engineering training was one of the factors that enabled the new industrial designers to discern what was wrong with the products of industry. Their common denominator was a strong aesthetic sense and all were greatly influenced by the Bauhaus.

For a long time there was a feeling in the industrial design profession that engineering training was more of a hindrance than an advantage. Some felt that anyone who understood the practical engineering side of a development could not appreciate the aesthetic aspects. That is why, in this book, I have tried to emphasize my training in the visual arts. Both my father and my mother were commercial artists. My brother, my sister and I all sketched and painted constantly as we grew up. My sister became a prolific author, my brother was a highly regarded avant-garde painter. These creative genes have repeated in the next generation. One of my sons is a leading architect in Colorado, another is a movie director with a series of feature films to his credit.

Actually, the link between engineering and art is well established, as is the link between art and mathematics. According to famous mathematician Norbert Weiner, aesthetics are an important part of mathematics; "...mathematics is essentially one of the arts," writes Weiner in his autobiography. "Neither the artist nor the mathematician may be able to tell you what constitutes the difference between a significant piece of work and an inflated trifle, but if he is never able to recognize this in his own heart, he is no mathematician and no artist." I have included a number of examples of the artist/engineer in this book, together with their accomplishments as well as some of my own, and I feel that the combination is more common than we realize.

By the same token, the ideal industrial designer should have an understanding of basic physical and scientific principles as well as an innate aesthetic sense. With no training or expertise in these areas, his or her work will always be confined to the embellishment of exteriors. Products such as the line of vacuum cleaners I did for Montgomery Ward, based on novel aerodynamic and construction principles, would never have happened without some engineering aptitude. In fact, the whole profession of industrial design is tending toward a more inventive approach. In the early days pioneers such as Loewy, Dreyfuss and my father were usually content to modify the exterior form of objects which were conceived and engineered by others. I have done plenty of this kind of work myself, but even in the 1930s I was starting to make innovative products from scratch, where I invented completely new engineering features, and a number of these became winners. Several ex-members of my staff have followed me in this direction and have become highly successful.

For the same reasons that some engineering experience is an asset to an industrial designer, it is an asset to an engineer to have some aesthetic aptitude and training. When I decided on a change of careers in 1942, and went to the Bendix Aviation Corporation as a junior engineer with no degree, I first had to catch up with thermodynamics, fluid flow formulae, and characteristics of materials. As I began to take responsibility for actual equipment development, I soon discovered that my holistic, intuitive approach to problems and the variety of my industrial design experience, gave me a decided advantage. I was able to arrive at simpler, more effective solutions, which proved to be functionally superior and less costly to build. Eventually this led to my becoming head of the Research Engineering Department, which was developed into a large, self-supporting organization. But my first love was industrial design and I left Bendix in 1952 to become a partner in my father's firm, Walter Dorwin Teague Associates.

Early industrial design firms operated on the principle that all credit for every job go to the head of the firm. I could never accept this and it led to constant friction. In some areas, such as interiors, my father was much better than I was, but in other areas—product design and automobile design—he was less competent. In the case of my own sons, I rejoice in their successes. My feeling was that having a son to succeed him who was expert in certain design areas would be a unique advantage to our company. But my father's generation of designers would not admit that anyone—even a son—could be equal, much less surpassing in any field of design. To me, the acknowledged authorship of outstanding accomplishments is more important than the monetary reward. This was one reason I had left industrial design to become an engineer at Bendix with a 50% reduction in salary.

The reader will find a lot of discussion of sports participation, including sailing, scuba diving, amateur automobile racing, outboard racing and skiing. These sports have been an important factor in my life: aside from the resulting physical fitness, they gave me a big advantage in the design of sports-related products, a large percentage of my work. Participation in sports creates an instinctive feel for the proper ergonomic design direction in non-sports products as well. It sharpens the ability to make fast, accurate decisions, an important business attribute. Finally, sports are a vital antidote to the stress of a heavy work schedule. I enjoy writing about sailing as well as design, witness my many articles in *Yachting*, *Saturday Review*, *New York* magazine and elsewhere.

In the course of years in design, I developed a technique we used successfully for product development. Having in mind both the schools that teach industrial design and practicing designers, I have included in Appendix A, a description of this technique, together with an actual example of how it worked in one case. Also included is a list of the U.S. patents in my name and a partial list of products of my design. For all products I have added the names of staff members who participated in the design process. My records and memory are pretty good for these projects, but not infallible. I apologize in advance for any involuntary errors or omissions.

I have enjoyed my life immensely, including the writing of this book. It has given me a chance to look back and review it all. I hope the reader will enjoy the book, too.

W. Dorwin Teague

Nyack, New York
September 1998

CONTENTS

1910–1920
CHAPTER ONE
Few of the residents had cars in 1915
when we moved in 10

1921–1928
CHAPTER TWO
Outside of the suburbs of San Francisco,
paved roads largely ceased until Kansas City 18

1928–1930
CHAPTER THREE
Uncle Widmer called Ellington's "The
Mooche" a "musical skunk" 24

1930–1933
CHAPTER FOUR
The Marmon Sixteen body design was
the chance of a lifetime 32

1934–1938
CHAPTER FIVE
More or less by common consent,
I became the chief product designer 48

1939–1941
CHAPTER SIX
I also learned to trust my instinct and
to question sacred dogma 62

1942–1944
CHAPTER SEVEN
The experienced engineer would be
too smart to even consider this kind of solution 74

1945–1947
CHAPTER EIGHT
He saw a big pool of red
fuming nitric acid seeping onto the floor 90

1948–1950
CHAPTER NINE
This was a useful lesson in
how not to build a rocket test stand 100

1951–1952
CHAPTER TEN
Every time Harry fell, Lewis
purposely took a tumble himself 112

1953–1957	**CHAPTER ELEVEN** The Count knew all the right places to go provided someone else picked up the check 118
1958–1959	**CHAPTER TWELVE** Clyde Cowan had the ability to explain recondite theories in understandable terms 132
1960–1962	**CHAPTER THIRTEEN** There is no way to tell a can of beans from a can of apple sauce 146
1963–1964	**CHAPTER FOURTEEN** This was flattering since Walt Disney had been making a strong pitch for the job 158
1965–1966	**CHAPTER FIFTEEN** At the Operakalleren the head waiter turned out to be a fellow Ellington buff 166
1967–1969	**CHAPTER SIXTEEN** The new firm's location was decided upon by finding the epicenter of the hometowns of the people involved 172
1970–1974	**CHAPTER SEVENTEEN** At the time the popular concept of conservation was one of discomfort and sacrifice 182
1975–1986	**CHAPTER EIGHTEEN** The variety of my work, if anything, increased as I approached my late seventies 196
1987–1998	**CHAPTER NINETEEN** "Dorwin, you're the one we want. Why can't you do it in the hospital?" 206
	EPILOGUE 216
	APPENDICES DTI Design Process; U.S. Patents; Life List of Projects; Report to Edsel Ford, et al. 220
	INDEX 250

CHAPTER ONE

1910 – 1920

Few of the residents had cars in 1915 when we moved in ...

I was watching the small plane as it circled low over the West Side Tennis Club grandstand and taking pictures of the National Championship match between Bill Tilden and "Little Bill" Johnson. Suddenly, without warning, the plane dove straight into the ground. I ran into the house, shouting the news to my mother, grabbed my bicycle and peddled madly to the crash site, a half-mile away. The pilot, Navy Lieutenant Grier, and the photographer, Army Sergeant Saxe, had been killed instantly. The plane had missed the grandstand and crashed in an open field nearby.

The year was 1920, I was ten years old, and my family had been living in our new home in Forest Hills Gardens since 1916.

There aren't any vacant lots in Forest Hills Gardens anymore, the trees are bigger, and it would be hard to see the plane over the grandstand today. But the basic character of the town hasn't changed much. If you blindfolded a stranger and released him in the middle of the Gardens, he would never guess he was in New York City. Despite being surrounded by massive apartments and urban sprawl, Forest Hills Gardens still has the character of a small country village. The tall flagpole, once the mast of the America's Cup defender *Columbia*, still stands in the village green, the West Side Tennis Club is as active as ever and the Forest Hills Gardens Corporation, representing the homeowners, still owns and maintains the streets, sewers, street lights and parks.

The character of the Gardens was created by architect Grosvenor Atterbury and landscape architect Frederick Law Olmstead. Atterbury set up the architectural restrictions which specified that all houses must be brick or stucco variations of the neo-Tudor style, with red tile roofs. Olmstead, of the famous Boston firm which designed Central Park,

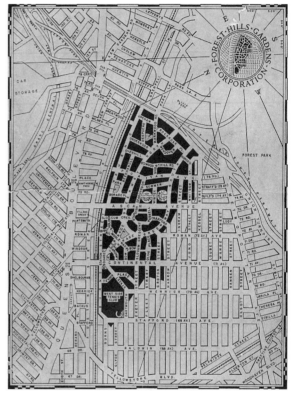

West Side Tennis Club. Forest Hills plat, 1915. The Teague home. A Sage Foundation house.

laid out the curving streets and little parks which gave Forest Hills Gardens its "cozy domestic character." Atterbury designed the Forest Hills Inn and Station Square complex, the art nouveau street lamps, the pebble concrete trash receptacles, and many of the early houses. Atterbury also invented a new system of factory-built housing. Several groups of these homes in the Gardens, erected as early as 1915, were factory prefabricated of concrete modules, pre-cast in the Sage Foundation factory on Burns Street and brought to the site by horse and wagon. My mother-in-law lived in one of these houses until 1970; it was as solid as the day it was built and showed every sign of lasting forever. Atterbury had a rare combination of innovative ability along with excellent aesthetic taste. In Forest Hills he conceived a daring plan for an original model community and invented radical building techniques that are still working successfully, eighty years later. Spending my early days in the midst of his work helped to give me appreciation for good design that has remained with me all my life.

Few of the residents had cars in 1915 when we moved in. Our house at 83 Beechknoll Road, like most others, had no driveway or garage. Ice was delivered every day or so by horse and wagon, and heating was by a coal furnace shaken down and cleared out by a "furnace man" who showed up early in the morning before anyone else was awake. Until oil burners began to replace coal around 1924, basements were dirty places where everything was covered with coal dust and ashes. We had originally planned to move in before 1916 but were held up by the casement windows, made in England, which were delayed by the onset of World War I.

I attended Public School No. 3, on the other side of the tracks, about a mile from my house, which was considered too close for busing. I walked each way winter and summer. The school was near Flushing Meadows, a large salt marsh full of wild fowl, rabbits, muskrats and foxes, later to be the site of two New York World's Fairs. Forest Hills was surrounded by farm land; one of my schoolmates was Mildred Vanderveer who lived on the nearby family farm settled by her Dutch ancestors in 1640. At that time the Meadows were criss-crossed by a series of wooden flumes which led the water from deep artesian wells to a steam-powered pumping station near Alley Pond where we used to skate in the winter. At a gun club on the outer edge of the Meadows we young boys used to hunt ducks and trap muskrats, usually with small success.

My sister, Cecily, was a year and half younger, my brother, Lewis, was born in 1917, shortly after we moved to Forest Hills. Both Cecily and I were born in Manhattan where we lived on Morningside Drive over the old Polo Grounds, original home of the New York Giants. My earliest recollection is of playing with a little cast iron toy racing car. Later we moved to Staten Island while our house was being finished. My mother, who was an ardent naturalist and amateur ornithologist, used to take me for walks to the salt marshes and boatyards near Great Kills. Automobiles, racing, sailing and salt marshes have been my major interests ever since.

Because of the West Side Tennis Club, Forest Hills was a great tennis town. When the Nationals or Davis Cup matches arrived, players from California and abroad were quartered in numerous private homes. One night a party for the visiting players, complete with orchestra, was held at a house near ours. My sister and I lay awake listening to the band which must have been an avant-garde combo because at one point they began playing "Limehouse Blues" and really swinging it. This was the first real jazz I'd ever heard and I was enthralled.

One of my friends was Joachim Nin, brother of Anais Nin. The Nins were Catholics and Joachim, in spite of his age (he must have been about eleven), played the organ for the services at the small wooden Catholic church on Ascan Avenue in Forest Hills. He had to go over there one Saturday to drop off some music so I went with him. Having heard about the strictness with which Catholics practiced their religion, I was astounded at Joachim's temerity when he switched on the organ and started to play. I was more horrified that he was cranking out a loud jazz tune but no flashes of lightning or irate priests appeared.

The United States finally got into the Great War, as it was called. In 1917 my father joined the 9th Coast Artillery, a National Guard unit assigned to guard the New York City water supply at Kensico Dam in Westchester. The "Forest Hills Bulletin" of August 1918 carries a mention of Lieutenant Teague. He finally made the rank of Captain; I still have his .45 caliber Colt automatic, which was issued to all the officers. One of the members of his company that he brought home was an American Indian named Amos Oneroad. The two of them were walking on a path near the Flushing Meadows one Sunday when Oneroad leaned down and picked up a flint arrowhead, the only one I ever heard of being found in the area.

Our family did not own a car until 1926. My father wasn't interested in automobiles and didn't learn to drive until around 1950. Riding with him then was a frightening experience as he occasionally lost concentration and wound up on the sidewalk or on the wrong side of the road. Fortunately he never had a serious accident. Once in a while my mother took me along on a New York shopping trip. Walking to Queens Boulevard, we caught the trolley, which eventually crossed the Queensboro Bridge and landed us near Bloomingdale's. The trolleys came along every ten minutes and the ride cost five cents. In those days before traffic jams and gridlock, the trip didn't take any longer than it does today in a car. My father had his office at 210 Madison Avenue, near 35th Street, and commuted on the Long Island Railroad.

During our first summer in Forest Hills, we rented a house in Ocean Beach on Fire Island. My memories of Fire Island are sand, wicker furniture, boardwalks and lots of poison ivy. Then we started taking summer vacations in a series of rented houses in Woodstock, New York, which was already an active artists' colony where my father had many friends. Getting up there entailed much preparation and packing of trunks. These trunks were taken by taxicab to the Day Line pier in Manhattan, where they were loaded on one of the paddle steamers— either the *Robert Fulton*, the *Henrik Hudson*, or the *Washington Irving*. The trip up the Hudson, with

Dorwin's maternal grandmother, Abbie Ann Fehon (née Smith). His mother Cecil (on the right) with her sisters Maude and Amy.

stops at 129th Street and various other towns before disembarking at Kingston Point, was always pleasant. Watching the massive paddle wheel machinery at work fascinated me. The Day Line was proud of the immaculate condition of its engines which were purposely arranged so that the machinery was in full view of the passengers. The *Robert Fulton* had a single huge vertical cylinder driving one end of a walking beam located above the boat deck. The other end had a connecting rod which was connected to a crank on the paddle wheel shaft. The other two boats were later models with three cylinders set almost horizontally. The cylinder on the *Robert Fulton* was six feet in diameter and had a stroke of twelve feet. The huge connecting rods and every nut, grease cup and fitting on the engines were highly polished steel or brass. The paddle wheel revolved relatively slowly and the horsepower, around 4,000 in the *Fulton*, was small compared to later turbine-powered vessels. But the effect created by the enormous mass of steel flying around gave an impression of irresistible power that no steam turbine or diesel could equal.

At Kingston, we were met by a large omnibus, drawn by a pair of draft horses, which carried us the last ten or twelve miles to Woodstock. Our neighbors there included Eugene Speicher, George Bellows and many other well-known artists. At the art gallery in the center of town I was constantly exposed to the paintings and lithographs of our neighbors which must have had an influence on my future. Woodstock was a great place for children.

We swam in the Saw Kill (kill meaning creek or brook in Dutch) at a place called the Little Deep, which was formed by huge, smooth sculptured rocks, carved like Henry Moore figures into various pools and round bathtubs by centuries of water and sand. The clear water was piped into many of the houses for drinking and washing. One house we rented, on the Ohayo Mountain Road, had a pump house located back near the Saw Kill with a primitive make-and-break engine connected to a pump that pumped the brook water up into a cistern in the attic. My duty, at the age of twelve, was to start this brute whenever the level of the water in the cistern got too low. As anyone who has ever worked with one can testify, starting a make-and-break engine is a miserable chore. One must let go of the compression release at exactly the right point while hand cranking, otherwise it will kick back violently and even start to run backwards. Downstream there was an abandoned saw mill which had a narrow gauge railroad, where we could ride one of the small cars down a long ramp. I built a rather elaborate model of the mill one summer, complete with rails from my standard gauge train set. I painted it red like the original with white trim around the doors and windows.

Both my mother and my father were commercial artists who met while attending the Art Students League. My mother did fashion drawings and continued to sketch and paint water colors, mostly landscapes and still lifes, until a few years before her death in 1960. One of three sisters who grew up in Maplewood, New Jersey, she was a descendant of

Edward Ball who settled what is now East Orange and Maplewood, with Robert Treat in 1666. Four of her great, great grandfathers fought in the American Revolution, two of them as captains. Mother used to take me over to Maplewood occasionally to visit relatives, a trip involving trains, ferrys and trolley cars. A lot of Revolutionary War activity took place around that part of New Jersey, General Washington used to visit the Timothy Ball family during the campaigns. Our relatives in Maplewood used to speak of the Revolution as if it had happened quite recently, which is understandable as there were people still alive then who had actually known Revolutionary War veterans. A distant relative, Uncle Sam, who visited us once in Forest Hills in his Packard Twin Six, used to dig up bullets, belt buckles and other battle souvenirs on his farm.

My father first came to New York from Pendleton, Indiana as a young man. His Teague grandfather had arrived in Indiana from North Carolina in a covered wagon in 1840 and his father was a Methodist "circuit-rider," a traveling minister who covered several congregations in different towns on horseback each Sunday. Because that wasn't a very lucrative occupation, he did tailoring part-time and later became a full-time tailor. He made the overcoat my father wore when he came to New York City in 1902. My father was born in 1883; by the time he was sixteen he had a job as handyman and reporter on the local paper. His schoolbooks are full of very neat notes and pen-and-ink sketches of elementary physics experiments. Arriving in New York with $35.00 in his pocket, he moved into the YMCA, where he checked hats and lettered signs to pay for his room and board and night classes at the Art Students League.

By 1908, after beginning at the Ben Hampton Advertising Agency, my father had a well-paying job with Calkins & Holden. In 1911 he rented a hall bedroom in an apartment house at 210 Madison Avenue, near 35th Street, where he became a free-lance advertising artist. His specialty was the fancy borders used on many ads in those days which came to be known as "Teague borders," even when some other artist did them. He was already making a good living and his income continued to increase, even during the Great Depression. Part of the reason was a friend who was a fellow inmate of the YMCA in 1902.

His name was Jack Brophy. In the early 1900s the two of them used to pool their funds and rent a canoe to paddle across the Hudson River and hike along the Palisades. When my father first started free-lancing he couldn't afford an accountant but by the early thirties, as the business expanded, he hired Jack as comptroller, which position he held even after my father died in 1960. Brophy wasn't perfect

but usually his advice turned out to be sound and had a lot to do with the growing success of my father's firm.

It is difficult to put the proper perspective on family finances of those early years. In the old Sage Foundation files, now located in the Rockefeller Archives in Pocantico Hills, I ran across a lengthy correspondence between two of the Foundation executives in which they argued as to whether a family with an annual income of $3,000 could afford to buy and keep up a house in Forest Hills Gardens. Their conclusion was that a single detached house was impossible, but if the family bought one of the attached row houses and exercised sufficient care, they might be able to make it.

In any event, my father was able to buy our Forest Hills home, which he helped to design with

Page opposite: Walter Dorwin Teague, 1902. Left: Cecil Teague with her sister Maude. The car is a 1907 Cadillac Model K Light Runabout. It weighed 1000 pounds, was powered by a one-cylinder engine of 98.2 cubic inches developing "over" ten horsepower, and was priced at $800.00 (the victoria top was a $125.00 accessory). The car belonged to Maude. Dorwin's father wasn't interested in automobiles. The Teague family would not own a car for another two decades— until 1926, a green Chrysler sedan that Dorwin picked out.

the architect, and to pay for his Manhattan office at the same time. A typical suburban dwelling of that era, the house had five bedrooms (four of them quite small), one bathroom, a generous living room, dining room, kitchen and sun porch, plus a full basement with furnace room, laundry and storage rooms. What distinguished it from most of today's housing was the solid, permanent construction mandated by the Forest Hills Gardens architectural restrictions. The basic structure was hollow tile with stucco outside and plastered walls inside, heavy red tile roof, and bronze casement windows with leaded panes, all of which made for a house which, barring fire or Force 8 earthquakes, should last indefinitely with minimum maintenance. Energy efficient, the hollow tile was a good insulator and the windows were relatively small and tight; there was enough mass in the walls and roof so that, as I remember, the house was never too uncomfortable in the summertime, even on the third floor. Another important factor was the construction by the Sage Foundation Homes Corporation, which built well, but unfortunately only broke even or actually lost money on many of the early Forest Hills Gardens houses. I drove by the house in early 1986 and it looked exactly the same except that an oak sapling I planted was now almost a foot in diameter and the owners have added a driveway and a two-car garage.

Inside, the house was furnished mostly with French and English antiques. Chippendale tables, good rugs and lots of books. We had a foot-powered Aeolian Hall player piano with a collection of music rolls that included both classics and some mild ragtime. My father's taste in interior design was excellent; later he did a number of interiors for clients like the Fords and his own apartment in the River House in New York City was featured in various magazines.

Almost from the time we could walk, my sister Cecily, my brother Lewis and I were artists. We all sketched constantly and painted with watercolors. Neither my father nor my mother ever tried to influence us or force us to follow any special kind of training or career. Rather they encouraged any kind of creative activity. As it turned out, my sister became a writer, my brother a painter, and I became a designer and inventor. All of us read a lot, our parents keeping us supplied with many good books. I remember best the large volumes illustrated so beautifully by N. C. Wyeth, including *Scottish Chiefs* by Jane Foster, *Kidnapped* by Robert Lewis Stevenson, *The White Company* by Conan Doyle, *Mysterious Island* by Jules Verne, and others. Howard Pyle was another great illustrator of children's books. These illustrations were as inspiring as the stories themselves and there has been nothing to equal them since.

While still quite young, I was given a simple set of Stanley woodworking tools: saw, hammer, chisels, etc. in a fitted wooden cabinet.

Both Walter Dorwin Teague and Cecil Teague were accomplished artists. The sketch could have been by either of them, but Dorwin believes that his mother was responsible for this lovely portrait.

Appropriating one of the basement rooms for a shop, I built various models and gadgets. Because of the coal furnace, the shop was usually dusty; I was also handicapped by a lack of instruction or experience. My father, oddly enough, had no interest in wood or metal working. While his pen and ink layouts required a certain amount of manual dexterity, he never applied this to any other medium such as painting or sculpture. On the other hand, I have always been fascinated by the inner workings of things like clocks, guns, cars and planes, as well as by their outward appearance. As I grew older, I became increasingly adept at wood and metal working: making things in my shop has been an effective form of therapy which brings me closer to the actual materials and how they worked. When I show some detail of construction in a drawing, I know exactly what series of actions are entailed to arrive at that result and consequently I am more likely to suggest designs which can be manufactured easily.

A few years after we moved to Forest Hills, Fred Stone, a well-known actor and musical comedy star, built a rather large house up the street from us. One of his successful musicals was *Stepping Stones* which also starred his daughter, Dorothy. She was older than Cecily but a younger sister, Paula, was Cecily's age; they became close friends and played in some amateur theatricals. Walter Hartwig headed an active players group in the Gardens. My father designed some of the stage sets and scenery for these shows. The December 1920 "Bulletin" described the "beautiful setting designed by Walter D. Teague for the play *Rachel*." The first Forest Hills movie theater wasn't built until 1922 but some early children's movies were occasionally shown at the Church in the Gardens.

Another Forest Hills resident was Fred Goudy who was probably the most famous typographer in the country. With my father, who was also a typographer, and Bruce Rogers, he formed the Pynson Printers. Father also belonged to the Salmagundi Club, a typographers' group, and the Advertising Club, where he took us to dinner occasionally.

The Fourth of July was a big celebration. Canvas covered the pavement in the station square and there were foot races and sack races for all ages. The big event of the 1917 Fourth was a speech by ex-President Teddy Roosevelt. I wasn't especially interested but my parents insisted that I go with them so that some day I could say that I had seen him. I don't remember a word of his message, but I'm glad I was there.

Wireless was just beginning to appear in 1917. A neighbor in a house behind us on Greenway North had a very professional rotary spark gap transmitter which was loud enough to be heard several blocks away without a receiver. Boys' magazines carried articles on how to make crystal sets. Most of these involved a "loose coupler" which was made by winding thin lacquered wire on a used Quaker Oats carton but the critical element was the little crystal on which a fine wire impinged. A pair of headphones and a long outdoor antenna with porcelain insulators completed the assembly. By fiddling with the coupler (the Quaker Oats carton) and tickling the crystal in the right place, one could occasionally pull in faint Morse code signals, far too fast for us to decode. As time went on, however, the early broadcasting stations, such as KDKA in Philadelphia, began to transmit and once in a while, when conditions were right, one could hear some music or speech. An affluent friend, Baldwin Vose, had an Atwater Kent radio at home with several large, black, graduated Bakelite dials on the front of the cabinet and a separate horn-shaped speaker like the ones used with the early Victrolas. These strictly-functional early radios were more beautiful than later models done by industrial designers.

A victim of infantile paralysis (polio), Baldwin wore a steel brace on one leg and was driven to and from school by a chauffeur in the family Cunningham. He occasionally gave me a lift. Once I distinguished myself by opening the door before the car came to a full stop. It contacted one of Grosvenor Atterbury's pebble concrete trash bins which gouged a triangular notch in the edge of the Cunningham's aluminum door. A heavy, expensive automobile usually sold with special bodies, the Cunningham was made in Rochester, New York. Later on we had a chap in the office named Stan Chamberlain who had worked there. He told me of one man who wanted a Cunningham with a coal-burning fireplace built into the back of the front seat structure. Attempts to dissuade him were fruitless, so the car was built and, as predicted, burned to the ground some months later.

As far as my childhood was concerned, I was lucky in many ways. We lived fifteen minutes from the heart of New York City with all its attractions, museums and theaters but at the same time we had plenty of open country and wildlife near at hand. The only negative factor was my rather poor relationship with my father which never improved very much as long as he was alive. I had a difficult personality myself so the problem wasn't one-sided. From his employees and children he demanded admiration and unquestioned obedience. Neither of these was I willing to concede, even as a child. All my friends seemed to have fathers with whom they could laugh and play together as they would with playmates of their own age. My father wasn't the playing type, moreover I was never able to talk freely with him. Many of our conversations quickly deteriorated into arguments.

This may sound like my father was a tyrant but in many ways he was very generous. We all received good schooling in private schools during a period when many families had trouble paying for rent and food. Father bailed me out of trouble numerous times when I was arrested for speeding or other offenses. He bought me cars and gave me a generous allowance while I was in college. He encouraged me to charge my clothing in expensive stores. In spite of this I can't remember many really happy times we shared with each other.

Chapter Two

1921 – 1928

Outside of the suburbs of San Francisco, paved roads largely ceased until Kansas City ...

In 1921, when I was eleven, my parents decided to enter me in the Kew Forest School, a small private school between Forest Hills and Kew Gardens which had opened three years earlier. The teaching staff was generally dedicated and capable. Most important to me, Guy Catlin, who taught math, was one of those rare teachers able to make his subject intensely interesting. He did not have a warm, outwardly friendly manner. Indeed I remember him as stern and reserved, but somehow he made his subject fascinating. The fact that in prep school I was able to get consistently high marks in algebra, trigonometry and geometry was largely due to Catlin's early conditioning. Another teacher, by persuasion and persistence, finally succeeded in making my handwriting fairly legible but, alas, it has regressed over the years. In five years at Kew Forest (skipping the seventh grade), I also played on the school soccer team and the football team, where I was best in defense.

When eleven, I was sent to Camp Wyanoke on Lake Winnepesaukee in New Hampshire for the summer. The lake was familiar to me because one of my uncles, Widmer Doremus, a Newark surgeon, had a cabin there which we used to visit. Being awakened each morning by bugle call was irksome at first when, at my uncle's a few miles away, I could get up when I chose and spend all day in a canoe if I felt like it. On the whole, though, I liked the camp, especially the annual excursions to Pinkham Notch where we visited the ice caves on Wildcat Mountain and rode up to the top of Mount Washington. I also liked the workshop at camp which was much better equipped than mine at home. There I designed and built a model of a contemporary speedboat which I sketched in one of my obligatory letters to home.

My grades at Kew Forest weren't too bad. I remember being in the finals of the spelling competition against Collier Elliott. But I was also a troublemaker and had to be disciplined quite often. Jack Gooding, my friend in high school, and I used to encourage each other in mischief which caused a lot of problems for the faculty. Matters finally came to a head in my junior year; at the end of the first semester the authorities called us in and announced that one of us, they didn't care which, would have to go. Because I was usually the ringleader, I was selected to leave and finished the year at the Richmond Hill Public High School which was academically quite easy after Kew Forest.

That summer, 1925, I had my first experience with the sea. A friend of my father's, Robert Adams, was an official in the Black Star shipping line whose 10,000-ton freighter, the *Santa Cecilia*, was scheduled to travel to the West Coast and back. Arrangements were made for me to make the passage on her. Bunking in one of the cabins of the officers' mess, I was listed as a supercargo, but actually worked the day shifts as an ordinary seaman. This consisted mainly of chipping paint, painting, or whatever menial task the bosun gave me. I was seasick for about ten days after we left New York but by the time we reached the Panama Canal I was beginning to enjoy the sea.

Passing the Canal was fascinating and the crew had time to see something of Balboa and Panama City. Our first port on the West Coast was San Francisco, where we had a few days while cargo was unloaded. When we got to Portland, Oregon, up the Columbia River, the skipper arranged for me to leave the ship and rejoin her in Seattle. I signed up with a group of twelve or so who were going to climb Mount Hood, 11,250 feet high, just outside Portland. The only boots I had were ten-inch, leather-soled hiking boots, so one of the guides inserted a number of small screws in the soles and heels, which stuck out about a quarter of an inch to form calks. We started out with several guides, climbed to the snow line and camped for the night in a shelter. At four the next morning, still dark, we started for the top, roped together in two groups. The idea was to get there and back before the sun melted the snow and ice enough to make it unsafe. Everyone was issued an Alpine stock to help us climb.

Ascending the steeper slopes, one of the guides led the way, chopping steps with his ice axe. Every once in a while chunks of ice and snow would break loose, finally one of them caught a guide squarely in the forehead. That finished him for the day; he had to be helped back to the base camp. Why more of the party didn't get hit I don't know. There was no tarrying at the top. Most of the party descended roped together, but several of us younger

Below: Dorwin's letter to his family from camp with sketch of boat he designed and built.
Right: His photos from St. Cecilia voyage, 1925.

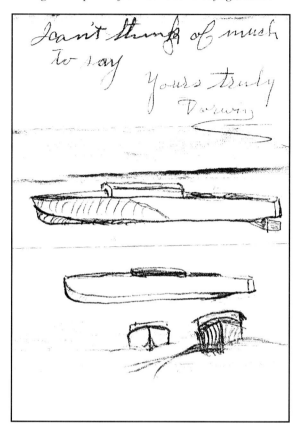

members elected to take the less vertical slopes by glissading on our rear ends, using the Alpine stocks as brakes.

Seattle still had a frontier air about it in 1925. Because of my interest in American Indians, I bought a small totem pole which looked crude enough to be authentic. Our last port was Vancouver, after which we departed for the Canal and home.

Before getting back to school I had a few days in New Hampshire on Lake Winnepesaukee where my family had rented a summer house. My friend Karl Easton was visiting so we decided to make a hiking trip from Pinkham Notch up to the top of Mount Washington and then across the Presidential Range to Randolph at the other end. The first day we climbed up through Tuckerman's Ravine to the top of Mount Washington and then descended to the Lake of the Clouds hut, a little below the summit, for the night. Everyone turned in early after dinner. About four o'clock the next morning we got up and dove into the frigid lake of the Clouds for a swim. Pitch black, dark and foggy with a cold wind blowing, somehow it seemed just the thing to do.

After breakfast Karl and I started out across the Presidential Range trail, which leads over the tops of Mount Washington, Jefferson, Adams and Madison, down to the little town of Randolph at the north end of the range. It was still cold and foggy, with visibility just good enough to spot the rock cairns which marked the trail.

The Presidential Range is not high by western standards, but it doesn't give away anything to the western mountains as far as weather is concerned. In fact some of the worst weather in the world has been recorded on the top of Mount Washington, including a wind velocity of 231 miles per hour in 1934. Even in July a violent snowstorm can make up rapidly with sub-freezing temperatures, high winds, snow and zero visibility. The upper slopes of the mountain are dotted with brass plates on the rocks marking the spot where some summer parties perished, in some cases within a few yards of a shelter they couldn't see.

After a while the wind let up but the fog persisted until we reached Mount Adams when suddenly, it lifted like someone raising a curtain, to reveal a view, clear as crystal, of New Hampshire and Vermont spread out below us in the sun. I always remember this as one of the most dramatic sights in my life.

The family rented a cottage on Sewall Road near Wolfboro for several summers. Every year we were visited by an Indian, with homemade baskets and moccasins for sale, who spoke the original Abenaki language better than English. I used to ask him to tell me the Indian names for things. He sometimes made me presents of rawhide and soft tanned deer skin which had a clean sweetish smell that I remember distinctly. One time I inquired about the meaning of the name Winnepesaukee which the Wolfboro Chamber of Commerce translated variously as "The Smile of the Great Spirit" or "Beautiful Water in a High Place." Actually, according to my Abenaki friend, the Indian pronunciation sounds more like "Elmeepesaukee" and translates to "Way Back Around the Bushes," referring to the main access trail in pre-highway days. Obviously this didn't quite fit the image that the Chamber of Commerce was trying to promote.

On my return to New York in the autumn of 1925, I was enrolled in the Cathedral School of St. Paul's in Garden City, Long Island. Compared to Richmond Hill High, this school was far enough advanced in Latin and math that I had to repeat my junior year of high school, which canceled out the seventh grade I had skipped at Kew Forest. Most of the pupils at St. Paul's were boarding at the school but a number of us—including Dick Earl, Rod Blackhurst, Ronnie Balcom, "Pickles" Heintz and me—commuted back and forth daily on the Long Island Railroad.

After catching up with my studies during my first year at St. Paul's, I began to get better grades. My only bad subject was Latin, due mainly to a lack of interest. This was a pity, as the little I retained was always useful in later life, helping me to understand other languages and to enjoy the classics. I also continued French, which I had already studied for several years at Kew Forest.

That same year I got my first airplane ride. A barnstorming pilot had an old Curtiss Oriole in a field near Garden City and was taking people up for a couple of dollars a ride. The Oriole was an antique, even in 1926; the engine was mounted in the open under the upper wing, above and behind the cockpits, where it would be sure to land in an accident. I didn't tell my parents about the flight until it was over.

The next summer, 1927, I signed up on another freighter, the *Charles H. Cramp*, of the Black Star line. She was also around 10,000 tons but with a flush deck rather than a well deck like the *Santa Cecilia*. Through some distant relatives in San Francisco I also arranged to team up with their son, Bud Strom, to buy a cheap car and drive back across the country.

The passage through the Canal to San Francisco was uneventful. My best friend on board was the radio operator. Ship communication was in Morse code in those days and the operator was always known as "Sparks." He wrote me when I got back to New York that, after I left, the ship grounded on the bar at the entrance to the Columbia River and had to be towed into Portland, where she spent the next few months in dry dock being repaired.

Bud Strom worked in a Sausalito garage and was a good mechanic. Without his experience we probably couldn't have made it across the country. Our car had to be reliable but inexpensive. We finally settled on a 1924 black Chevrolet roadster that we covered with clever sayings in white paint, a campus fad in those days. In 1927 California had no driver's license or insurance requirements so, with California plates, we were entirely legal at age sixteen. Outside of the suburbs of San Francisco, paved roads largely ceased until we reached Kansas City; most of the Lincoln Highway, the best and most direct transcontinental route, was mud, sand or sharp gravel. In hot weather the gravel raised havoc with our 30 by 3-1/2 high pressure tires. Experienced drivers warned us about this before we left, so we started with several spare tires lashed alongside. Before journey's end three more tires had to be bought plus innumerable tubes. Tire changing in those days was not just a matter of unscrewing or tightening a few lug nuts. The tire had to be pried off the rim, the tube removed and patched, and

The cross-country Chevy, before being festooned with slogans, Bud asleep on the spare tire, his sister and Dorwin.

pumped up again after it was replaced.

Thanks to Bud's know-how we had few mechanical problems. The only recurring trouble was with the clutch which was of the leather-faced cone type. After use, the leather would become glazed and the clutch would start to slip. Fortunately it was exposed to the open air beneath the car, so the fix consisted of having one person work the clutch pedal in and out, while his partner lay under the car, flipping sand up onto the clutch face with a large tablespoon kept in a side pocket for the purpose.

After various adventures we reached Pendleton, Indiana, populated largely by my father's friends and relatives and some of Bud's family friends. We received the royal treatment as transcontinental heroes, huge meals and dates with the town's prettiest daughters. Quite low in funds by now, Bud and I decided to earn some money by picking tomatoes on a large local farm. Two days of this proved enough to disenchant us with agricultural work so we wired home for money. Because Bud decided to stay on a little longer with the good life in Pendleton, I drove the last 700 miles to Forest Hills alone. The Chevy was the first car in the Teague family.

In 1926, while I was traveling to the West Coast on the *Santa Cecilia*, my father took a vacation in Europe where he visited England, France, Germany and Italy. He was especially impressed by what the Bauhaus people were doing and realized the important implications for the United States. The more he thought about this after he returned, the more convinced he became that he should give up advertising art and become an industrial designer.

Exactly when the title "industrial designer" came into being is obscure, but it was already in use by 1930. Its origin derived from the lack of attention to appearance in many new machine-made products. Older, long established goods—i.e., dinnerware, furniture and clothing—had their own specialized groups of designer/artists. But more recently introduced products like telephones, typewriters and appliances were designed, engineered and sold with little thought to appearance or even human factors in many cases. These virgin fields were lumped under the heading of "industry." Ergo, someone whose business it was to redesign these items for better appearance and useability became a "designer for industry" or an industrial designer.

My father's first major industrial client was the Eastman Kodak Company in 1928. Eastman had been looking for a designer or artist to improve the appearance of its line and Adolph Stuber, the chief executive officer at that time, asked the people at the Metropolitan Museum for advice. They recommended my father. This was the start of a relationship that lasted thirty-two years, until my father died in 1960. His first assignment was to redesign the leather carrying case for existing camera models and then to design decorative appliqués for the existing camera itself. Later he designed the new Baby Brownie. Previously the Brownie line had consisted of rectangular box cameras covered with imitation leather. The Baby Brownie was to be a box camera but more compact, molded in a thermosetting plastic, very simple and inexpensive. My father designed it with rounded corners, in brown phenolic, with a series of vertical ribs around the periphery. The new camera was an instant success. Eastman not only sold a lot of Baby Brownies but, even more importantly, loads of film to go with them. The company was delighted.

My father had stipulated a very substantial fee for the times. This was a smart move as it not only insured that his opinion would be respected, but it set the pattern for subsequent Eastman assignments and for dealing with other clients. With the success of the Baby Brownie, the future of my father's industrial design business was assured.

As my sister Cecily and I reached our teens, we were allowed to go to the local movie house or, as a special treat, to attend a Broadway show. The Forest Hills movie theatre, located on Continental Avenue, had opened in November of 1922. Silent picture theatres usually had a rather elaborate organ with an organist who varied the musical mood to suit the action on the screen. There was little subtlety. Even blindfolded, one knew when things were going well for the heroine or when the villain was sneaking up on her.

My first Broadway show was *The Cocoanuts* with the Marx Brothers, which opened at the Lyric Theatre on 42nd Street in late 1925. Although the brothers had appeared on the New York stage earlier, this was their first real musical comedy. George Kaufman, assisted by Morris Ryskind, wrote the script which added up to an irresistible combination; the music was by Irving Berlin. For my sister and me the show was love at first sight. Alexander Woolcott's opening night review said it best: "I cannot recall ever having laughed more helplessly, more flagrantly and more continuously in the theatre than I did at the way these Marxes carried on last night … *The Cocoanuts* is so funny that it's positively weakening."

Animal Crackers, which followed in 1928 at the Forty-Fourth Street Theatre, was another Kaufman and Ryskind blockbuster—even funnier than *The Cocoanuts*. As in all Marx Brothers' shows, the action was punctuated with outrageous puns by Groucho and Chico, which lesser comedians probably couldn't have gotten away with. About the time one became exhausted from laughing, Harpo would sit down and perform one of his truly beautiful harp solos, or Chico would play the piano. Even while Chico was clowning and "shooting" the keys, there was no mistaking that one was listening to a virtuoso.

One of Harpo's many standard silent gags was to start leering at some attractive chorus girl until, getting more and more nervous, she would begin running away frantically with Harpo in close pursuit. Some years later my sister told of waiting for someone in the Commodore Hotel lobby when who should come in but Harpo. My sister was quite attractive; Harpo spotted her immediately and went into his act. She said it took all her will power to refrain from jumping up and dashing for the exit, because she knew Harpo would be right after her.

Another memorable event for me was the first time I saw Bill Robinson perform. The show was *Blackbirds* of 1928 or 1929, I think, an all-black review on Broadway. Robinson, known best by his nickname "Bojangles," was regarded as the best tap dancer of all time. His style was based on keeping a single, steady rhythm, not too fast and not too slow, from start to finish. If an electronic oscillograph recorder had been available in those days, the readout would have shown only milli-second variations from the basic beat. In the thirties, tap dancing was popular in vaudeville and musical comedies. Most tap dancers included a lot of gymnastics in their routines which were admittedly difficult to execute but usually required a slow-down of the background accompaniment. This

The Baby Brownie. Photo courtesy © Eastman Kodak Company. Page opposite: "Bojangles," in studio portrait and circa 1930 sketch by Dorwin.

always spoiled the act for me and I was annoyed that the audience applauded loudly for these tortured maneuvers.

Robinson was best seen when he could be heard. The ideal accompaniment was either none at all or, at most, a very soft piano or orchestration which took the beat from Robinson, rather than vice-versa. In some of Robinson's later appearances, such as the film *Stormy Weather*, the orchestra overpowered Bill, so that the taps couldn't be heard distinctly, very frustrating to the aficionado. Also, I think that audience reaction was important to Robinson, and this was lacking in his movie appearances.

One of Robinson's inventions, his only prop actually, was a flight of three or four steps in which he constantly seemed to come up with new combinations. It's difficult to image how a dancer, with a flight of steps and a derby hat, could hold an audience spellbound for a half hour at a time. Robinson did it with a dazzling variety of steps, all performed in an effortless manner, while never deviating from the steady mesmerizing beat. According to one account, at a dancing teachers' convention in 1932, he danced an hour and five minutes without stopping or repeating a step. In *The New Yorker* (October 8, 1928 issue) he answered a question about the secret of his dancing with "I dance close to the ground," which is about as good a way of putting it as any.

During my senior year at St. Paul's, I was still commuting on the Long Island Railroad and getting good marks except in Latin. It was time to think about college. Because of my liking for math, geometry and trigonometry, I finally decided to try for MIT. With long commuting hours and plenty of homework, I no longer had time for sports but, although I had grown considerably, I was still too light for St. Paul's college-weight football team anyway. At the end of the year, I took the final College Board exams, doing reasonably well in English and French and getting 100 in trigonometry, 100 in plane geometry and 95 in solid (the result of a minor error). The only subject I flunked was Latin. I was accepted for MIT, providing I could make up the Latin during the summer. I graduated from St. Paul's in June.

So during the summer of 1928, while the rest of the family was up in New Hampshire, I was home trying to conquer Latin with Dan Patch who, according to local legend, was the "second best Latin teacher in the United States." He hammered it into me. On my second attempt I achieved a 65 in Latin in the College Board, just enough to squeak by. Earlier in the year, I had disposed of the Chevy and talked the family into buying its first car, a second-hand Chrysler in good condition. It never got much use since I was the only one in the family who could drive. When it came time to go up to Cambridge to register, my father gave me a new Chevy roadster as a reward for graduating from St. Paul's.

Chapter Three

1928 – 1930

Uncle Widmer called Ellington's "The Mooche" a "musical skunk" ...

In the fall of 1928 I packed my gear and headed for Boston to enter MIT. Hank Carlton, a six-foot-four freshman I met during registration, and I decided to take a room together on Marlborough Street. With my St. Paul's math background, the first semester was quite easy and my marks were good. The fraternity rush got underway; Hank and I were pledged by Phi Sigma Kappa so we moved into the fraternity house. Freshman pledges were subject to a certain amount of hazing and doing the less attractive jobs in the house. For a big dance in November, our room was selected as the girls' cloakroom and we were ordered to clean it thoroughly for the occasion. We did this but I had an idea that would make for a little more interest: arranging everything in the room on a very slight slant. The bureaus, tables and chairs were propped up on the left side, the curtains and pictures adjusted so that they all leaned about two or three degrees to the right. This created a startling drunken effect as one entered the room—even the upper classmen were sufficiently intrigued that we escaped serious punishment.

The freshman differential calculus course was especially interesting because the instructor was Professor Norbert Wiener who received his bachelor's degree at age fourteen, his doctorate from Harvard at nineteen, and who was one of the great mathematicians of all time. He and Vanevar Bush were jointly responsible for developing one of the first computers. At the slightest sign of any interest by one of his students in some unusual problem, Professor Wiener would happily cover the blackboard with the formulae to demonstrate the mathematical proof of the question. One Christmas Eve Wiener and Bush visited a friend in Cambridge to help him set up the electric trains he had bought for his son. In spite of all that brain power, the locomotive stubbornly refused to move. After various futile efforts someone finally remembered that the apartment was in a section of Cambridge still supplied with DC current.

MIT had no varsity football team at that time, but I did go out for the freshman team, which was a fairly informal organization. We played some prep school varsity squads and other college freshman teams but seldom won a game. As the winter set in, I got on the swimming team which had a high percentage of Phi Sig members. MIT had no pool so we had to practice in the Boston University Club pool and, of course, all our games were away. We competed with Amherst, Brown, Yale, Navy, Columbia, etc. and even won a few meets. My specialty was the 50-yard freestyle and the relay.

Most girls' colleges then had a strict curfew and were very particular about their students' companions and where they went on dates. This tended to cramp our style. Inadvertently, Jack Rogers avenged our constraints once. He owned one of the first L-29 Cord automobiles, very low and exotic looking, although mechanically somewhat anemic. One evening while driving over to see his girl at Smith, Jack approached an ancient cantilever bridge just wide enough to accommodate two cars. He was late so he was pushing it a bit, unfortunately lost it, and crashed into one of the main girders which knocked the end of the bridge off its underpinnings. No one was hurt but the bridge was partially collapsed. All Smith girls and their dates behind Jack were forced to take a tortuous twenty-mile detour so the sacred curfew was missed by quite a few. After being untangled from the girders the Cord was driven away. The bridge had to be closed for repairs. The Auburn Automobile Company had photos taken of the wrecked bridge with the slightly damaged Cord stuck in it and circulated the picture widely to demonstrate the rugged qualities of its car.

Prohibition, which lasted until 1933, had little effect on drinking, if anything it made liquor more exciting. There was no problem at all in getting whatever one wanted; depending on the quality, the price was anything from a few dollars a gallon up. The cheapest moonshine was available in South Boston, reputedly distilled from garbage. Many fraternity houses made their own bathtub gin. Every student was entitled to purchase a quart of pure alcohol for chemistry experiments but many didn't use it so the system was to buy up as much as possible and mix it with distilled water, oil of juniper and other ingredients. I expect the quality was at least as good as much of the gin sold today. "Near beer," which was very slightly alcoholic, was legalized and could be fortified with alcohol to produce quite potent "needle beer."

Auburn Automobile Company photo used to demonstrate the rugged durability of its new L-29 Cord. Photograph from the R. C. Greene Collection. Left: Teague family outing on Mirror Lake. Dorwin's mother is at center in a hat, his sister Cecily at left with her Forest Hills sportsman beau.

Making home brew was also a popular pastime. This involved mixing water, malt, raisins, etc. in a ten-gallon ceramic crock. After a week or so, the beer was decanted into quart bottles and capped. The hardest part of the process was waiting for the sediment in the bottles to collect on the bottom so that the beer was ready to drink. Once we made and capped about thirty-five quarts and, when the level in the crock decreased, discovered a large cake of yellow laundry soap which a rival faction in the fraternity house had slipped into the batch while it was fermenting. After some debate, the decision was to finish bottling and see what would happen. Waiting awhile, we ultimately gave a bottle to a member of the group that we suspected as the perpetrators, saying it was from another batch. He drank it with no apparent ill effects, in fact he praised the quality, so we drank the rest ourselves.

Liquor was everywhere, if you knew where to find it. Whenever I had a date with a girl at Smith, I

stopped off at a hotel in Worcester to warm up. (In an open car, winter drives were brisk, to say the least.) To get to the speakeasy there, one went into the men's room and rapped on the tile wall past the end urinal. The proper signal opened up a long old-fashioned mahogany bar, complete with brass rail.

I was drinking a lot. But so were most of my fraternity brothers. I didn't think it was a problem. My schoolwork did begin to suffer. Still, the generous funds I received from home did nothing to lessen my appetite for high living.

Fred and Peyton Cooper were brothers from Dallas who also joined Phi Sigma Kappa. The three of us decided to move out of the fraternity house and take an apartment together in the neighborhood. Both Coopers had a good musical background. Their favorite instrument was the banjo. In addition to jazz and blues they could play semi-classical stuff, each taking different parts. During their freshman year a Boston movie theater invited local colleges to supplement the regular stage show and vaudeville on selected nights. I suggested Fred and Peyton should represent MIT. They agreed instantly and turned out to be the big hit of the whole program. The audience clapped, whistled and cheered, made them play encore after encore, and the orchestra leader offered them a permanent job with the band. We also had a quartet, in which I sang the lead, Fred took tenor, Peyton was baritone, and Hank Carlton bass. Our specialty was spirituals and old blues numbers like "St. James Infirmary." We sang at a number of parties including a big affair my father gave at his New York office.

College students were much dressier in those days. We wore suits with jackets to class, fedora hats and dress overcoats outside in the winter. Parties brought out dinner jackets and for special affairs we wore white tie and tails. My father treated me to a set of tails when I was a freshman, plus all the accessories—white silk scarf, white kid gloves, patent leather shoes, opera hat and silver cigarette case. At one dance I remember vividly, in the ballroom at the Copley Plaza, Duke Ellington's orchestra was playing and I became an avid life-long fan. An orchestra member—I think it was Johnny Hodges—gave me a reefer, as joints or marijuana cigarettes were called then. Fortunately I couldn't detect much of any effect, immediately lost interest and was never tempted to try drugs again.

Ellington's "The Mooche," issued on the Okeh label in 1928, was the first of my collection of jazz recordings which would include all of Ellington plus Louis Armstrong, McKinney's Cotton Pickers, Jellie Roll Morton, King Oliver, Basie, Cab Calloway and others. The 78 rpm records were easily breakable so with frequent moves from fraternity house to apartment and back, attrition was high. That summer—1928—I took "The Mooche" up to Lake Winnepesaukee where Uncle Widmer had weekly classical music evenings at his camp on the lake. "The Mooche" horrified him; he called it a "musical skunk." The recording remains one of my favorites despite the vast improvement since in high fidelity techniques.

Voo Doo was the student comic magazine at MIT—for which I did several cartoons. One of our main peeves was the local laundry, a truly miserable operation that sent our shirts back regularly with collars frayed, buttons missing and heavy starch in spite of instructions to the contrary. My cartoon was a full-page diagram of the inner workings of the laundry, including collar sandpapering machines, acid vats, button-picking machines, etc.

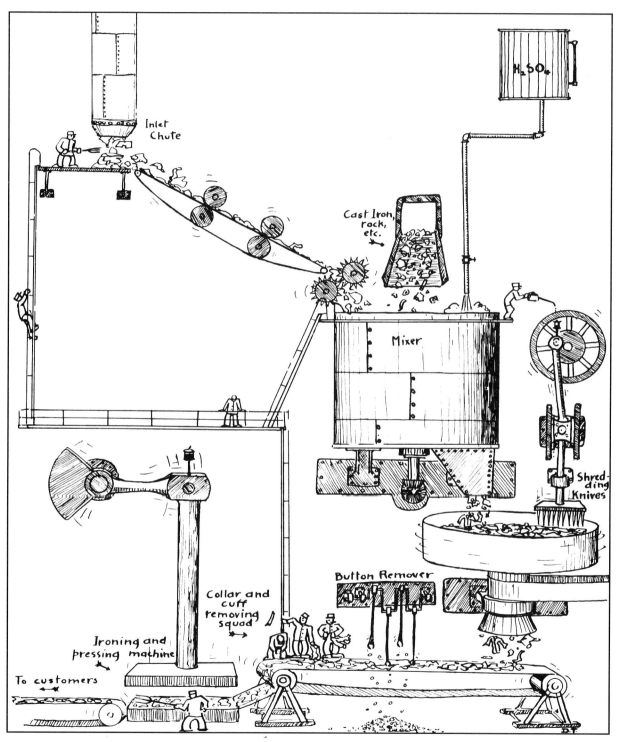

Page opposite, below: Faded photograph from college days: snappily-dressed Phi Sigma Kappa fraternity brothers on their way to school, Dorwin second from the left. Page opposite, above: college friend Fred Cooper in a photo from his World War II Air Force days that he sent to Dorwin as a postcard in September of 1943. Above: Dorwin's take on the local laundry. The cartoon appeared in the October 1929 issue of Voo Doo and was roundly applauded by his fraternity brothers.

For summer vacation, my father arranged a job for me at the Eastman Kodak plant in Rochester. Because I had taken a mechanical drafting course at MIT and had some experience at home, I was able to make myself useful although the job wasn't especially inspiring. The cameras usually underwent a number of changes during the year as problems showed up or less costly ways to build them were devised. These changes had to be made on the drawings for the particular parts and that was my job. Every morning when I arrived for work, a large stack of old ink drawings would be placed on my desk with the required change orders. If I managed to finish them during the day, another stack would appear. My pay during that summer was $25.00 a week.

At the YMCA, my first home in Rochester, I met a friend my age from Nashville and we decided to share rooms in a boarding house. A place on Meig Street was located, quite comfortable, with good food, for ten dollars a week each. Rochester was a quiet, pleasant town, relatively cool in the summer. On weekends we canoed on the Genesee River and the Erie Barge Canal which runs right through the town. With luck we could hitch a tow behind a slow-moving barge, lie back and watch the scenery go by. Although I was not receiving any allowance from home, by exercising some care I was able to save up enough before the summer ended to buy an old Harley-Davidson motorcycle. Within an hour after I bought it I was back to have the handlebars straightened and get a new headlight, having learned about trolley car tracks the hard way. Oddly enough, though rough riding, this old machine turned out to be rugged and reliable transportation. But it was heavy, with a high center of gravity, so picking it up after a spill required real effort.

Because the Eastman job hadn't been very creative or exciting, I wasn't sorry to leave Rochester after my tour of duty was up. I packed up my few belongings, lashed them to the pillion seat of the Harley and took off for Lake Winnepesaukee 450 miles away. No matter how tightly I lashed my baggage on, it kept falling off. The roads were as rough as the bike, with lots of sharp corners and various road blocks in the form of hay wagons, cows, etc. The only brake was on the rear wheel, not a very efficient arrangement. I ran out of fuel with only enough money to call home. The family Chrysler, driven by an English friend of my father's, came to the rescue and towed me to the Brick House, our new place on the lake.

That same year my father had looked at the old Piper farmhouse bordering on Lake Winnepesaukee. The house was brick, built around 1810, and located on the main road, just north of the little village of Mirror Lake—close to the property of Uncle Widmer. As old snapshots show, the farmhouse was primitive but basically sound with a large barn and ramshackle sheds. The property was quite beautiful with a long open hay field leading down to the lake, a family graveyard in the middle of the field, and the ubiquitous New Hampshire stone walls. Although a bit out of true, the framework of the barn was of sturdy hand-adzed beams, morticed together and secured with wooden tree nails. Heavy chestnut boards nearly three feet wide separated the stalls. One of the carpenters told me that there wasn't a tree left in New Hampshire that would yield such boards.

My father saw the possibilities of the property and, after much consultation, decided to make an offer. Ralph Piper accepted, a price one percent of the property's value today. At first my family camped out in the house which was unfinished on the second floor. The first order of business was to put in a bathroom, enlarge the kitchen and finish the upper floor rooms. A long dormer window on the back side of the roof provided more space for the top floor rooms and bath. Dismantling the makeshift chimney gave us fireplaces in the parlor and sitting rooms. The ramshackle wood shed was removed entirely.

The next project was the barn restoration. Before and after photos show the exterior changes. The new plan called for a large garage at the left, then a "shop" section which I designed with a double door and a ramp on the north facing the lake so boats and cars could be brought in and out, a large skylight over the doors and a raised balcony

Page opposite: Dorwin on his Harley, 1929. Above and below. The Piper house and barn before and after restoration.

Above and below: The barn restoration. Right, from the top: Dorwin discovered an ancient highwheeler in the barn and had to try it out. His mother at a New Hampshire auction. The barn interior as decorated by Cecil Teague. Page opposite: Brother Lewis in the outboard hydroplane Dorwin designed and built.

overlooking the work floor. Next was the rather spectacular living room, 30 by 50 feet with a ceiling height of 25 feet under the peak of the roof. The original frame, made of hand-adzed timbers, morticed into the uprights and secured with wooden pegs or tree nails, was kept intact except for the lower horizontal course which provided insufficient headroom. These members had to be raised a couple of feet which called for some expert craftsmanship by the carpenters. A large fireplace and massive mantle was built at the east end of the living room with a flight of stairs leading up to a comfortable guest room and bath over the new alcoved shed openings. The entire barn, roof and sides were shingled over.

One reason for the success of this renovation was the timing, right after the onset of the Great Depression when jobs and money were scarce. New Hampshire always had plenty of good craftsmen and the crew working on the Brick House and the barn were the cream of the crop. No task was too difficult. I was up there while the work was in progress and I used to like to eavesdrop as they traded New England type witticisms and insults.

There were many auctions in New Hampshire in those hard times. The auctioneers all knew my mother well. She was expert at spotting valuable pieces under layers of pink paint. Gradually the whole house and barn were filled with antique furniture and bric-a-brac selected to fit special spots. Years later, when my sister owned the house, tragedy struck. During the winter when there was little traffic on the lake road and nearby houses were empty, a gang of thieves hid their moving van down by the lake and used my sister's pick-up to shuttle the loot there. Everything was cleared out except the giant folio Audubons on the walls of the barn which, curiously, the thieves missed.

My first use of the new shop was to design and build a very simple stepless outboard hydroplane for which my long-suffering Uncle Widmer let us use his 4 hp Johnson outboard. We removed the exhaust plate which increased the noise, I'm sure the other residents on our bay were relieved when my uncle finally decided he had better restrict his motor to fishing while it was still intact. I also owned one of the original Old Town sixteen-foot canoes. These were canvas covered over steamed wood planks and ribs, and were made in Old Town, Maine where many of the employees were American Indians. This canoe is surely one of the classic designs of all times.

Chapter Four

1930 – 1933

The Marmon Sixteen body design was the chance of a lifetime ...

During my second year at MIT my father got a new client, Colonel Howard Marmon of the Marmon Motor Car Company. Marmon was located in Indianapolis and had a reputation for building fast, high quality cars, incorporating advanced engineering. The first Indianapolis 500 in 1911 was won by the Marmon Wasp driven by Ray Harroun, using a rear-view mirror, a new invention. I don't know how my father first met Colonel Marmon but they became good friends and remained so until Marmon's death in 1943. Tracing back their family histories, they found both had grandfathers who migrated from the South in covered wagons and settled in Indiana in the same year and possibly in the same expedition. Also both ancestors were millwrights, the equivalent of a mechanical engineer or machinist today.

The first work my father did for Colonel Marmon was to make color suggestions for some of the models preceding the Sixteen. These were pretty startling compared to the rather drab choices available at that time and Colonel Marmon must have liked them because he approached my father about designing the bodies for the expensive new sixteen-cylinder car he was planning to produce.

I kept in close touch with what was going on in my father's office at 210 Madison Avenue and this news was exciting. Cars had been my first love since I was very young and at nineteen I had already covered lots of paper and the inside of school books with automobile designs. At sixteen I had driven my first car across the United States. Since around 1927 I had a subscription to the British *Autocar* magazine, the equivalent of today's *Road & Track*, which covered Grand Prix racing, road tests of new cars, the latest engineering innovations and European body design. I knew the names of all the top European and American racing drivers and their cars and I was thoroughly acquainted with European and American automobiles, especially the better quality examples like the Bentley, Invicta, Bugatti, Renault, Rolls, Isotta Fraschini, Hispano-Suiza and Duesenberg, Auburn and Stutz in the United States. Up until this time, my father had never been interested in automobiles and knew nothing about the mechanical features of their engines, chassis or bodies. So when Colonel Marmon asked him if he would like to tackle the body design for the sixteen-cylinder chassis, I put up some strong arguments for doing the design job—and my father finally agreed.

The rule in all the early design offices was that every design was attributed to the head of the office, regardless of who actually created it. The Teague office was no exception. The excuse given for this practice was that "the clients are paying for my services so my name must be on it, otherwise the client will think he is being cheated." The staff of the design offices used to joke about this in private. One prominent designer had a reputation for calling a meeting and saying, "I had an idea for the so-and-so account while shaving this morning," when everyone on the staff knew the idea had actually been proposed earlier by one of them. There was nothing the employees could do about it, since it was the reputation and the publicity surrounding the three or four top designers that brought in the business and any young designer who wanted to work in the profession had to go along with the system.

I knew instinctively that the Marmon Sixteen body design was the chance of a lifetime, and I knew I could design a better-looking car than anyone else. But I also knew that unless I spoke up early, I would get little or no credit and my part in it would never be known. Accordingly, I convinced my father that I could design a car that would be outstanding, but said that I felt that I should get at least equal credit for the design. He agreed but put nothing to that effect in writing. As it turned out, the only time he ever honored this agreement was in his book, *Design This Day*, in which the captions under the Marmon Sixteen, the rear-engine car rendering, and the Marmon Twelve, list me as co-designer. But later he assumed full credit.

I was nineteen when Colonel Marmon gave us the go-ahead, at least for a rendering which would show what the design might look like. This took place during my second year at MIT so for the first part of the job I commuted back and forth between Boston and New York on the train or, when the weather was good, on Colonial Airlines which was the only air carrier operating on that route in 1929. The planes were tri-motor Fords and the passengers received a certificate stating that they had actually

One of the perks of working for an Indianapolis automobile company, a visit to the Indianapolis 500 while in town on business, this view of the '29 race taken by Dorwin from the Henry Ford box.

flown the whole distance. The tickets also carried a clause in which the passenger agreed not to sue Colonial for any injuries that might be incurred Later in June, I came down to New York and worked full-time for the summer in the office at 210 Madison. By that time, my father had taken more space by knocking down the walls in two adjoining apartments, and had a staff which consisted of John Moss and a Miss Pritchard working for him.

Colonel Marmon had the chassis and engine pretty well designed by the time I started so I had definite dimensions to work with. I did the initial concept in the form of 1/10th scale drawings as there was no doubt in my mind whatsoever as to the design I wanted. In later years I rather turned away from the large, luxury cars in favor of small sports cars but in these early days the 145-inch Marmon Sixteen chassis was close to my ideal. I then drew a perspective of the car and made an airbrush rendering to give a better idea of how it would actually look. John Moss, who was an expert with it taught me how to use the air brush so the rendering turned out quite well. There were no trial sketches or alternate concepts. These scale drawings

and the rendering, from which I made the full-size drawings, were the first and only drawings I made. Aside from some minor details which I will describe, no one else, including my father, made any sketches, drawings, renderings or design concepts of any kind. The only exception was a little sketch which my father made as a design suggestion, in which the fenders, in side elevation, were angled in an octagonal fashion. It didn't take long to talk him out of this.

The Marmon Sixteen body design was primarily an aesthetic exercise with very little innovation other than the omission of the radiator cap and the fender skirt which hid the axles and brake mechanism in the front elevation, plus the partial integration of the trunk into the body which was new, at least for America. The design was rather a rearrangement of features, most of which had been used before, into a harmonious ensemble with general proportions, curves of fenders and roof line, removal of most of the usual excrescences, and the continuous belt line, that all together made for a beautiful, classic design that thrilled me then and still does today. The effect is best epitomized in the close-coupled five-passenger sedan, which was the first model I designed, and the convertible sedan. I also designed closed and convertible coupes. There was an attractive "victoria"-sort of a two-plus-two arrangement—but I'm not sure whether I designed it or whether it was adapted by someone at the factory.

The classic character of the Marmon Sixteen was the end result of years of study and sketching which had built up an image in my mind of how the ideal automobile should look. I knew precisely the curves and proportions of the fenders, the roof line and all the other dimensions and contours that I wanted; it was just a question of sitting down at the drafting board and putting it on paper. In the first 1/10th scale drawings a variation of 1/64th of an inch in the fender line would have ruined the effect that I wanted and it was lucky that I was later able to make the full-size drawings so that the chances for error in the translation to the three-dimensional prototype were minimized.

This all took place during my father's "octagon period" when he was fascinated by the eight-sided shapes and insisted on applying them to everything he designed. This wasn't so bad for some things but for an automobile it was a disaster. His other early clients were Taylor Instrument and Bausch & Lomb—both in Rochester, New York. Taylor made dial thermometers and barometers and he designed these with octagonal dials instead of round. I remember a scathing review of these designs by Lewis Mumford in *The New Yorker*, in which he characterized my father's barometers as "beautiful until he tried to make them so."

While it wasn't hard to convince my father that octagonal fenders would make the car look like a 1908 American Underslung, he did insist on applying his pet motif to a few other places. This included the hub cap emblems which were small and relatively harmless, but also the longitudinal junction of the top and sides of the hood and radiator. I couldn't persuade him that a radius was better here and was overruled. The scale model and the original full-sized prototype were built with this bevel but somehow it was then quietly changed to the radius I wanted in the first place. The only other exterior vestige of the octagon motif was the half-inch bevel around the front edge of the radiator shell which is rather nice and crisp, and the hub cap emblems.

During the summer of 1930 I also made the full-sized lofting drawings for the car. The office at 210 Madison Avenue had one wall big enough for the job. I used wrapping paper which was the only paper I could seem to locate around New York that was long and strong enough without a lot of patching together. Lofting drawings which include sections in all three planes have been used for centuries to define boat and ship hulls, and I had done such drawings for boats. I don't know if anyone else has ever used this method for cars but if you know what you want it goes fairly fast and it worked very well in this case. The first mock-up and the production dies were made directly from these full-sized drawings.

Boucher, a well-known model maker in New York built a 1/6th scale model of the car. This rather odd scale was chosen to fit the miniature 600x20 rubber tires on the Firestone glass ashtrays which were quite common in restaurants and speakeasys. Since duplicating the tread on six tires for the model would have been a tedious job, this trick made sense. The drawings for the Boucher model were prepared from my drawings at the Marmon factory in Indianapolis. The Boucher company did its usual meticulous job of duplicating every detail including the six wire wheels.

The only part of the interior I worked on, aside from the seating dimensions and steering wheel position, was the instrument panel. In 1930, the universal practice in America was to make instruments any shape but round and then combine them in an unpleasantly shaped cluster. Dial faces were usually beige or some other hard-to-read color. I designed round instruments, each one separate with black faces and white numerals, located in a row high up on the instrument panel so that each was easy to read. The rest of the interior hardware, such as the door pulls and handles, were designed by my father in the octagonal motif.

One reason that the Sixteen turned out as well as it did was because of the way Colonel Marmon operated. Strictly from the economic standpoint, the car was a disaster. A Marmon marketing department never would have sanctioned the idea of bringing out a high-priced, sixteen-cylinder automobile as the country was moving into depression. But Colonel Marmon probably wouldn't have listened anyway; this was his company and he ran it the way he wanted to. It was a rare privilege to work for such a person. Once he saw that a designer knew what he was about, he backed him up all the way. I remember that there were loud objections from the engineers when the radiator filler cap was removed from its traditional position on top of the radiator shell. He merely told them to "figure out how to do it their way." After a while, the engineers came to the conclusion that the filler could be concealed under the hood, where it remains in every car today.

When the first handmade prototype was finished, Colonel Marmon loaned us the car for several weeks with a chauffeur so my father could use it. The Sixteen went into production in April of 1931 and about 500 were sold all together. A surprising percentage are still running, most of them in

Prototype of the Marmon Sixteen Convertible Sedan, shot in Marmon's Indianapolis factory. The photograph above is Dorwin's favorite of all those taken of the Sixteen. Unfortunately, Marmon later had it extensively retouched, dubbing in a radiator ornament and ruining the roofline in the process of eliminating the factory background.

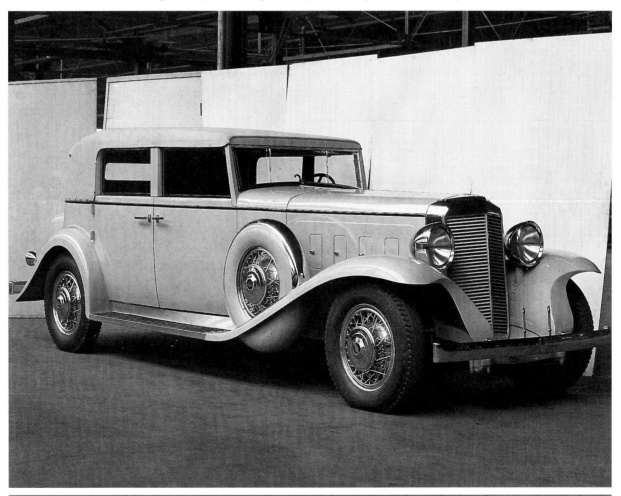

Below: The Marmon Sixteen Victoria. Page opposite: The Marmon showroom on Broadway in New York, 1931. Left: The first handmade prototype of the Marmon Sixteen as dispatched to Forest Hills for a visit with the Teagues, complete with chauffeur so Dorwin's father could use it. The man who designed the car's body is seen with it here.

impeccably restored condition, worth many times what they cost new. The Marmon Motor Car Company never developed into a Chrysler or even an American Motors, but it is a good thing for the industry that there have always been a few daring idealists like E. L. Cord, Ettore Bugatti, Ferdinand Porsche, Enzo Ferrari and Howard Marmon, who have put innovation, quality, performance and beauty ahead of maximizing profits. Certainly from my personal standpoint, the opportunity to design the Marmon bodies was the sort of assignment most designers dream of but never realize. For a nineteen-year-old boy to get this sort of responsibility was a lucky combination of circumstances that probably will never happen again. I have a fancy plaque from the Classic Car Club of America which reads, "The CCCA…hereby presents to W. Dorwin Teague Honorary Membership with all the attendant rights and privileges thereof, and with the hope that he will find as much enjoyment in it as the members of the Club have found in his work."

All the commuting to New York and extracurricular work on the Marmon didn't do much for my scholastic record at MIT which was, and still is, the sort of place that requires not only aptitude but dedication and hard work if one wants to graduate. There was no way I could do both so the school work suffered. I repeated some courses but the motivation just wasn't there and eventually the school decided I would be better off elsewhere. In retrospect I regret that I never graduated from MIT, but I did get enough basic training to allow me to take on responsible engineering assignments in later years. Certainly if I had passed up the chance to design the car bodies in order to apply myself sufficiently to graduate, it would have been the wrong decision. Thousands of engineering students graduate every year, but there aren't many nineteen-year-old automobile body designers.

After leaving MIT I formed an informal partnership with a friend who had a small machine and woodworking shop in his home on Beacon Hill. There I made some gunstocks, probably the most difficult and exacting of all woodworking tasks, and also some furniture. For fun, I designed a car that I had been thinking of for some time and then built a 1/10th scale model of it in the shop. As far as I know the pontoon fenders on this model had never been used before on a passenger car. I got the idea from the streamlined Stutz that had been built to break the world speed record at Daytona Beach and in which Frank Lockhart lost control and was killed

Below and right: Dorwin's "just for fun" model that he designed and built in Boston in 1931. Above: As adapted for the Marmon Twelve, this model seen before painting.

during one of the high speed runs. The Stutz fenders had been designed for purely aerodynamic reasons but they turned out to be very attractive on a passenger car.

The body of my model was built up of a number of wooden sections, screwed and glued together. A mahogany pattern for the radiator shell with the proper allowances for shrinkage was cast in bronze. I filed and buffed it to a high polish and then etched two brass plates to represent louvers and soldered them into each side of the casting. The conical disc wheels and tires were turned out in one piece from a four-inch brass bar on the lathe and the hubs and discs polished, leaving the "tire" portion rough. I also turned out the headlights, taillights and door handles on the lathe and had a jeweler fit a pair of watch crystals into the headlights to simulate the lenses. All the hardware was bright chrome-plated; the tires were painted matte black. The model was unpainted for a while and looked very attractive in this form. In 1932, after spray painting the body light gray acrylic, I mounted the model on a beautiful slab of rosewood I had been saving, and etched a little German silver nameplate for it with my name and date.

This body design was quite revolutionary for its time. My father asked if he could keep the model in his New York office. I agreed, but with the stipulation that it should be presented as my design. While I don't believe he actually told visitors that he had personally designed it, they certainly left with this impression, which he did nothing to correct. Later, in his profile for *The New Yorker* and other articles, he claimed full credit for the concept by which he had "revolutionized the industry." At any rate Colonel Marmon visited the office while I was back up in Boston and immediately fell in love with the model. At that time Marmon was testing a new twelve-cylinder chassis which had some very advanced and innovative features. The frame consisted of a single 8-1/2 inch diameter steel tube running down the center line, which split into a fork forward to receive the engine. Pairs of lateral quarter elliptic springs extended from each side of the tube, front and rear, to create four-wheel independent suspension and the body was to be mounted in three

Below: Dorwin's rendering for the Marmon Twelve. Page opposite: One of his 1/10th scale drawings fro the prototype of the Marmon Twelve. Dimensions were based on Marmon factory drawings of the chassis which consisted of an 8.5-inch tube split in front of the engine mount and sets of four cantilever springs in front and rear for four wheel independent suspension. The prototype itself, Dorwin's Woodlites replaced by Pierce-Arrow-inspired headlamps by Marmon.

points on the chassis. The engine was a V-12 with a lightweight aluminum block and stainless steel liners as in the sixteen. Colonel Marmon asked my father if he could adapt my model design to a body for the twelve-cylinder prototype.

I came down to the office to make the drawings and Marmon sent a set of chassis layouts. I made 1/10th scale side, plan, front and rear elevations for a close-coupled sedan. Modifying Marmon's steering wheel and radiator location slightly and indicating the seating dimensions, I showed an alternate version which had removable panels in the fenders for additional stowage spaces. Unfortunately there was no chance to make up more detailed drawings, or to follow up with the necessary modifications to fit the chassis, and the changes were done at the Marmon factory. This was a pity, as the prototype could have been a much more striking and attractive automobile.

Another airbrush rendering I made of the twelve showed a pair of octagonal bumpers my father insisted on. Fortunately these were omitted on the prototype in favor of some innocuous conventional types. Another feature shown on the rendering was a pair of built-in Woodlites, set into each side of the radiator shell. These headlights were a contemporary add-on item reputed to deliver bright, even illumination with very little upward spill over, although I never actually verified this. They were also eliminated on the prototype which I think was for the better, but instead of substituting free-standing headlights as used on my model, the factory or someone decided on a pair of Pierce-Arrow clones, built into the tops of the fenders in an ugly manner and then added triple ornaments on the radiator and the lights. The radiator grille is too straight up and down, the window sills are too high, and a pair of Pep Boys horns complete the desecration. Fortunately, my original model still exists to show what might have been. Some time back I donated it to the Cooper-Hewitt Museum in New York, where it can be seen along with the Boucher model of the Sixteen.

While it was frustrating to be denied full

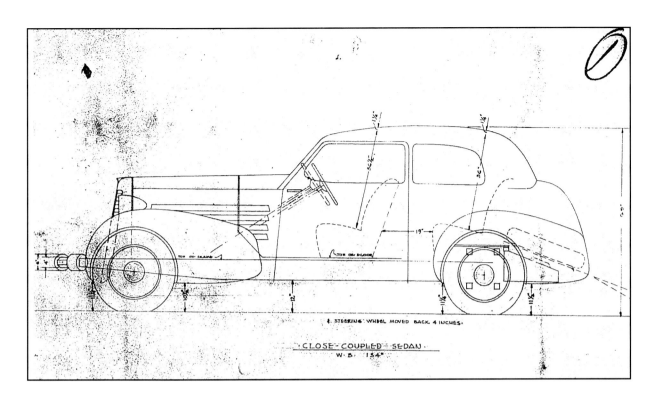

CLOSE-COUPLED SEDAN
W.B. 134"

recognition for the car designs, it was better than not having the opportunity to do them at all. Without my father's effective salesmanship and his friendship with Colonel Marmon, all this never would have happened. Lucky for me too was the timing in the early thirties. Nowadays, with hundreds of designers working for each automobile company, the one-man design is a thing of the past. I was also fortunate in that I had a thorough knowledge of automobiles and trends in automotive design, some background in aerodynamics and, more important of all, an instinctive feel for good lines and forms.

My father got interested in streamlining for cars, as exemplified by the "teardrop" shape, and he also became convinced that all cars would someday have the engine in the rear. Actually one rear-engined production automobile existed in the thirties. This was the Czechoslovakian Tatra. My father asked me to design a rear-engined car; I drew up three versions in 1/10th scale, and then made airbrush renderings of them. While these designs would obviously have a low drag coefficient, I've never been too enthusiastic about them. For one thing, rear vision was very difficult to achieve. Also, the cars would be aerodynamically unstable at high speeds, with all the bulk located forward. A long wheelbase permitted a fairly efficient aerodynamic shape, but if the design had to be adapted to a more practical length, the teardrop would be too blunt, separation would take place, and the shape would no longer be efficient. Finally, there is the same objection that many people have to today's vans: the front-seat passengers are too vulnerable in an accident. The German aerodynamicist Wunibald Kamm has shown that in most cases a sharp cut-off at the rear of the car is more efficient than a fully-streamlined teardrop, unless the latter is excessively long. It is true that Volkswagen had many successful years out of the rear-engined Beetle, but without a very good drag coefficient because it was too short. Of these objections the most serious is the lethal instability at speeds of 60 mph or over, which cannot be corrected even with a monster stabilizing fin. Rear-engined passenger cars have nearly disappeared for these reasons.

Certainly my drinking during this period contributed, with a lack of interest, to my failure to manage passing grades at MIT. But in New York, faced with the task of car design, I could control my consumption sufficiently to get the job done. I still didn't think I had a problem.

Page opposite: Rear three-quarter view of the Marmon Twelve. Above and below: Dorwin's rear-engined cars.

Above: The Teague family Auburn 8-120 Convertible Sedan, Dorwin poses, sister Cecily vamps, their Sealyham terrier begs.
Below: Lem Ladd's "Old Gray Mare."
Page opposite: Lewis and "The Mooche."

In 1931 during one of my visits to New York I talked the family into buying an Auburn 8-120 Convertible Sedan, built by E. L. Cord in Indiana. Since no one else could drive it, I took the car up to Boston with me where I continued to live except during most summers which I spent at the Brick House in New Hampshire. The Auburn was handsome except when the top was lowered to form a bulky, clumsy-looking lump at the rear of the tonneau. To solve this I removed the top entirely, stowed it in the barn and had a small canvas cover installed in its place. This made a very sporting-looking car but with no weather protection. At the time I considered this a minor disadvantage as I had plenty of warm clothes and if I went fast enough the rain would carry over my head. On each trip between Boston and the Brick House I used to try to lower the elapsed time record. On one of these trips I was doing about 80 miles an hour on the narrow lake road, only a few miles from my goal at the Brick House, when the right rear wheel broke off at the axle.

There was a confusing few seconds while I must instinctively have tried to maintain control. The mark left by the drum showed that the car had swerved to both sides of the road until finally coming to a stop almost two-tenths of a mile further on. The wheel had struck an approaching Chevy pickup so hard that it bent the Chevy's frame out of line. Some neighboring children were searching for it in the roadside woods where they had seen it bounce. How I missed the truck I'll never know; this was only one of my close shaves but I've never injured myself or anyone else.

While at school I had owned a LaSalle Phaeton for a short time and later, during a period when I was relatively low in funds, a huge old Stearns-Knight. A smooth, silent old relic with a beautiful gearbox, it had a voracious appetite for oil, thanks to the sleeve-valve engine. Later I traded this for a more modern Harley-Davidson motorcycle. Its only accident was on the Fenway. I had the right-of-way but a girl in a sedan didn't see me and pulled out past the stop sign just as I had gunned it to cross the intersection. Jamming on the brake (still on the rear only) I skidded hard enough to roll the tire off the rear rim so that the tube bulged out, hit the car and bent the forks. The only personal damage was a large rip in the seat of my trousers which was immodest but not painful. I was once arrested and fined in that same Harley for driving on one of the footpaths in the Public Gardens while standing on the seat.

One of my friends in Boston was George Wood who belonged to my fraternity and whose doctor father had been given one of the attractive new 1931 Ford Two-Door Phaetons. I helped him install special cylinder heads, pistons and carburetors until it was capable of an honest 100 mph, quite an accomplishment for a Model A. A dirt track race was coming up at the Rockingham, New Hampshire speedway, so George decided to enter. We sawed off the rear half of the body at the door sills, removed the fenders and lashed a drawing board in between the front seats. Dirt track cars were the usual at Rockingham so there was some opposition to our outfit but amazingly George was allowed to enter, mainly to amuse the spectators, I think. The announcer made the most of the comic aspect until George's qualifying time came out close to the top drivers. In the final event, George was actually running in second place for a good part of the race until a rod bearing let go in the overworked engine. I took 16 mm movies of the race which I still have.

In addition to George, I had several friends in the Phi Gam house in the Fenway including Jack Stover, Ed Marshall, George Harvey and Chuck Schauer. Our common interests were girls, Ellington music, P.G. Wodehouse and motor vehicles. We often used to visit the Oak Hill Garage in the suburbs, run by a racing enthusiast called Lem Ladd. Among the cars Lem owned was a built-up sports/racer known in pre-World War II days as the "Old Gray Mare" that he raced in the ARCA (Automobile Racing Club of America), the pioneer sports car club in the U.S. He used to let us work on our cars in his garage on Saturdays and there was usually something interesting going on. Lem was almost always among the top finishers in ARCA events and for many years he held the record on the Mount Washington toll road hill climb, one of the main ARCA trials. The Oak Hill Garage specialized in Lancias which were very advanced for this time with four-wheel independent suspension, four-wheel brakes and a low monocoque body/chassis. Jack Stover in our group had a Wills-Sainte-Claire Touring, one of the first U.S. production cars with a double overhead camshaft and Chuck Schauer, who later became executive vice president of the Research Corporation of the Foundation for Science, owned a new four-cylinder Henderson cycle which was the BMW of American machines. One night Chuck decided to see how fast he could lap the Fenway, a circular road that looped around the Back Bay Fens. A Boston motorcycle cop took off after him with siren screaming. With the wind and noise of the engine, Chuck couldn't hear the siren so he made several laps, almost lapped the cop and had plenty of time to hide the Henderson behind the Phi Gam house before the lawman caught up.

In 1932 my brother Lewis went to the Gow School near Buffalo. In its woodworking shop he built the frame and bottom for a step hydroplane outboard racing boat. I finished the boat in my shop in Forest Hills; we called it "The Mooche" after the Ellington record. The frame and sides were oak so the boat was strong but quite heavy for the motor which my father bought him. We both raced "The Mooche" in the New England outboard circuit without winning any silver but we had a chance to get to know the other racers and find out something about the sport. Outboard racing at that time was a low key sport and didn't require a lot of capital.

In between trips to Forest Hills or Lake Winnepesaukee, I lived in Boston or Cambridge in a series of rooms and apartments. Boston always had a number of good eating places. Our favorites were Durgin Park and Jake Wirths on Thursdays for the bratwurst and dark beer. For late evenings, the place I liked best was down by the New Haven tracks in South Boston: the Railroad Club. This was a black nightclub which stayed open late and served fairly potent needlebeer. The house piano player, George Tynes, was very good. Whenever the top black bands like Louis Armstrong and Cab Calloway were playing in Boston they came to the Railroad Club after their regular engagement to sit in. George was a wonderful accompanist and he liked the way I sang so, after I had a few beers to work up enough nerve, he would talk me into doing a few numbers with him. I sang McKinney's songs like "Three Little Words" and "If I Could be With You," but my specialities were "Minnie the Moocher" and "Smoky Joe." The patrons were mixed black and white and they would all join in the "Ho de Ho" and scat choruses. George later left the Railroad Club to join McKinney's Cotton Pickers as their regular piano player. I heard later that one of the band cars had a bad accident and George went through the windshield, cutting all the tendons in both hands. That may have been the finish of the band because it dropped out of sight; it was one of the three or four best at its peak.

During New York visits I used to see a lot of my future brother-in-law Arthur Meurer who was one of the world's greatest storytellers. He made his living in the airplane business, buying and selling planes and he had an early pilot's license. One day in 1932 he called me in Forest Hills to say he had borrowed an Amilcar two-seater sports car from a friend and would I like to go for a ride. I accepted at once. This was my first experience with a European sports car. The Amilcar was a rather primitive little staggered two-seater with wire wheels, cycle fenders and a high-revving engine. This one was painted white and had no top but the weather was fine so we decided to visit some of the better country clubs along the North Shore of Long Island. Nassau and Suffolk counties were not yet developed; most of the land was either in farms or in big estates. Jericho Turnpike and the Vanderbilt Motor Parkway were the main east-west routes which were narrow and winding, through beautiful country, just made for a sports car.

Almost all the country clubs were hurting, as many of their members had been forced to resign as the depression deepened. Driving up to a club Artie, who was a great salesman, would hint that we were looking around for a club to join. As two Fitzgeraldish characters in a fancy new white sports car, we were invariably invited in to look the place over, stay for a while and have a drink or lunch. This worked so well that we toured around for two days and I had a chance to find out what a joy it was to drive a car that handled well. On the second afternoon the Amilcar began to exhibit some expensive noises, in the form of a rod bearing knock, so we decided to take it back to the owner, Ray Gilhooley, in New York City.

Gilhooley's was one of the first, if not the first, sports car sales establishments in the country. He was an ex-race car driver whose chief claim to fame was performing a 360 in full view of the grandstand at Indianapolis and calmly continuing the race. This unplanned maneuver came to be known as a "Gilhooley" for many years afterwards. When we brought the Amilcar into Gilhooley's the showroom was full of exotic sports cars including a pair of Type 37 Grand Prix Bugattis. One was black and the other was fire engine red with a little sculptured devil thumbing his nose on the radiator cap. Both were apparently in brand-new condition. I had seen and admired pictures of Bugattis in *Autocar*, but the reality was overwhelming.

Until this time big cars were my ideal and I regarded body design as a primarily aesthetic exercise, separate from the mechanical functions of the car. Here in the Bugattis, it was immediately apparent that I was looking at a blend of function and aesthetics, put together by a master. No wonder, because Ettore Bugatti, *Le Patron*, had originally been trained as an artist and came from an artistic family. His father, Carlo, was a prolific furniture designer and builder, his brother Rembrandt, a renowned sculptor. Ettore had no engineering degree but he was one of the most prolific designer/engineers that ever lived. He held over 500 (some say over 1,000) patents and, although the Bugatti factory produced less than 10,000 cars altogether, there were 49 different models. These include the Bugatti Royale—the largest car ever built—and the Type 52—an electrically-driven child-size replica of the Grand Prix car.

Bugatti was an individualist who paid little attention to the way other car makers operated; he designed and built his cars his own way. The horseshoe-shaped radiator, the hollow front axle, the cast aluminum spoked wheels with integral brake drums, the excessive front wheel camber were only a few of the typical exclusive Bugatti trademarks. The damascened engines were works of art and, in the case of the Grand Prix cars, the whole machine, including the body, was a statement of functional beauty. One model of the Grand Prix car won 1,045 races in two years and one of the most beautiful passenger bodies, the Type 55 two-seater, was designed by Jean Bugatti, Ettore's son,

who was killed testing one of the cars at age thirty.

Bugattis engender a feeling of loyalty and devotion in their owners that no other man-made object has equaled. The saying "once a Bugattiste, always a Bugattiste" is at least true in my own case. Here was a piece of machinery that combined function and aesthetics in a way that was an inspiration for me for the rest of my life. I swore to myself that someday I would own a Bugatti. It took me another twenty years to achieve this, but from that day on I was strictly a sports car buff and no longer had much use for big cars.

In between periods of design I continued my high living in Boston. While dating various Boston girls, I began getting serious about Rita Lynn. My "playboy" life was beginning to bore me so I decided it was time to settle down. Rita and I got married and I took a job with my father's firm on Madison Avenue in New York. We leased an apartment on Austin Street in Forest Hills and I commuted to New York on the Long Island Railroad.

Top: An Amilcar of the type that made Dorwin a sports car enthusiast. Above and below: Two among the glorious lifetime of the marque from Molsheim: the largest and the smallest Bugattis. Ettore at the wheel of the fabulous Type 41 Royale. The Type 52 "Bébé" Bugatti, exquisite miniature of the Grand Prix car, bought by Dorwin's brother, Lewis with Dorwin's grandson August behind the wheel. Photo: Harry Teague.

Chapter Five

1934 – 1938

More or less by common consent, I became the chief product designer ...

By 1934 my father had several new accounts and was beginning to add to the staff. One new client was Ford, marking the start of a long association. Another was the New York, New Haven & Hartford Railroad for the design of a new passenger coach. An experienced railroad engineer was added to the staff to insure that the design would be practical. Several hundred of these cars were produced and were in use for many years afterwards. The interiors were simple and attractive; the exterior was improved by flush windows with rounded corners and the addition of skirts that came down nearly to track level. The job was well underway when I joined the firm and I had very little to do with it.

More or less by common consent, I became the chief product designer. Among the first tasks I tackled were the housings for a line of heaters, boilers and humidifiers for the Bryant Heater Company. I devised a standard method of construction for the line with radiused corner members to which the various side and top panels were spot welded, and then finished in dark blue crackle enamel to conceal the weld marks. From my detailed drawings nine pieces of equipment went directly into production and were sold for many years.

Another interesting job was the casing for a 600 hp radial natural gas-powered engine, driving a compressor and a generator for the Dresser Manufacturing Company of which Neil Mallon was president. An unusual assignment, it involved improved appearance and human factors for a piece of machinery used as a booster pump in natural gas transmission lines, where it was never seen except by company maintenance personnel. The original engine was very complex, encrusted with all sorts of pipes and ancillary equipment, and almost impossible to keep clean. Mallon's idea was that the less complicated and better looking the equipment, the more likely that it would be maintained properly and kept clean. He turned out to be correct, with the added bonus that the cover also reduced the ambient noise level in the room. The redesign was quite successful and marked one of the first cases of which I am aware that attention was paid to the exterior appearance of industrial equipment never seen by the general public.

At the opposite end of the spectrum was a piece of sculpture I did that same year for the Steuben Division of the Corning Glass Company. I've forgotten what the other designs were but my contribution was the rather supercilious stylized lion. I don't know if he was ever done in crystal but I hope he was. An extensive redesign of the Model 96 mimeograph machine for the A. B. Dick Company was next and, under John Moss' direction, I did the graphics for a line of mimeograph packaging. This was the first time I had worked in graphics and lettering.

I also worked on several other products during my first year at the office and, in reviewing these designs of fifty years ago, the volume and variety of work is striking. I completed a number of sizeable product design assignments from concept to final drawings which not only defined the final appearance but in most cases I supplied accurate dimensional mechanical details of parts and controls.

In 1934 industrial design was still in its infancy. No established design directions or styles existed; each product was being tackled for the first time. Many designers submitted a series of studies to the head of the office and then to the client. I never did this. After making some free-hand sketches with details of various features and methods of construction, I went directly to the final design. I don't remember my father disliking any of my concepts although he would sometimes try to get me to add some decorative feature which I usually resisted, sometimes successfully. My major criteria were cleanness, simplicity and functional compatibility, both as to the ergonomics and to the requirements of the particular manufacturer's production techniques. Occasionally the client's production people would question some feature that might create manufacturing problems and I would either modify the design or suggest a way to get around the problem. This didn't happen too often as I usually got a chance to become familiar with each client's manufacturing methods beforehand, and I had already developed a fairly extensive knowledge of materials and fabrication.

Much of the success of a particular assignment depended upon the personality of our contact in the client's organization. This was usually the chairman, president or the chief executive officer. His attitude

The High-Speed Radial Gas Engine for Dresser. Model of "Lion" for Steuben Division of Corning Glass.

would be reflected by the engineering and marketing personnel and could range from enthusiastic cooperation to suspicion and outright hostility. In the early years, cheerful acceptance of the designer's contribution was the norm; later on we would occasionally run into antagonism. In most cases I was able to convince the engineering people that I was sincerely trying to work with them and was not threatening their jobs. Nothing infuriates an engineer more than to have some outsider second-guess a product that is the result of years of hard work on his part. But if he can be persuaded that a better final result will be achieved, and that he will receive at least part of the credit for the improvement, an atmosphere of mutual respect and cooperation develops, which guarantees successful conclusion.

In later years one member of our organization, the account executive on some large accounts, would deal strictly with the person in charge of the client's organization. He determined first what this individual was expecting and then slanted the design in that direction. In meetings with the client he took the side of the top executive and denigrated any contrary suggestions from lower echelon people, regardless of the merit of their ideas. This worked well for him in many cases but sometimes his sponsor retired or left the company, and one of the staff that he had put down previously assumed

the top position. Our relationship with the company would come to an abrupt halt.

The client's marketing group was sometimes more of a problem in getting a new design accepted than the engineering department. With marketing we were dealing with intangibles difficult to define and pitting our recommendations against their years of experience in the industry. I'm sure every designer has encountered the complaint that, "We tried that back in 1954 and it caused nothing but trouble" or "the so-and-so company did that in 1962 and almost lost its shirt." These experiences may indeed have been bad news at the time but in the intervening years the public's taste could have changed radically or new materials and techniques been developed which eliminated the original cause of the trouble. Sometimes there was a reluctance to go to a radically new design because "it is ahead of its time." A favorite example used to justify this attitude was the Chrysler Airflow which failed, not because it was ahead of its time, but because it was a supremely ugly design. Usually the final outcome depended on whether the executive who was our contact had sufficient confidence in our recommendations and our past results to convince the marketing people that we were on the right track.

A few times I ran into complaints because I had only submitted one concept. Why hadn't I submitted several ideas so that they could make a choice? But I had automatically considered and rejected alternatives and had no interest in submitting anything but what I thought was the best approach. If the client remained adamant and insisted on seeing alternate ideas, my course of action was always to bow out of the relationship as gracefully as possible. Fortunately there have always been enough understanding clients, so that I have been able to concentrate on designs that I can still be proud of.

One area where I've always remained rather neutral is color. Color is such a subjective thing that I distrust attempts to generalize on its psychological effects. Statements like "yellow is manic" are foolish; in my opinion, individual color preference is strictly a matter of association. A person may like a certain color for his automobile, another color for room interiors and a third color for clothes. More sheer baloney has been written on color than almost any other subject. There are trends which are important to recognize although it is virtually impossible to assign any reason to them. For example around 1966 Porsche came out with what they called Bahama Yellow, sort of a pumpkin color. Because the new 911 Porsche had a well-deserved image for speed and handling, the color became very popular (I had one) and was used later on VWs and on some American cars. Nowadays it has almost disappeared. Some of the original Lamborghini Miuras were painted in a light zinc chromate—sort of a greenish yellow—that was cheerfully accepted by the buyers of this outrageous and beautiful machine. But it would have been unsaleable on any ordinary car. Fire engine red has always been so popular for sports cars that law enforcement officers inevitably associate it with exceeding the speed limit. At the time of this writing, black had almost completely superseded frost chrome plate in cameras and hi-fi equipment but we can be sure that some other finish will be in favor a few years from now. For the most definitive article on color, I recommend the piece by Edwin (Polaroid) Land in the May 1959 *Scientific American*.

Walter Dorwin Teague Associates' biggest job in 1935 was the design of a line of standardized service stations for Texaco. Bob Harper was in charge of the project. This was the first time an oil company opted for a standard gas station design. Over 40,000 stations were built worldwide to this design, which featured three green stripes running around the top of the building walls and the edge of the cornice over the pumps. My only connection with the project was a cosmetic exterior treatment for the Texaco fuel trucks. In 1935 the office also became involved in the design of exhibits and sales rooms for Ford, A. B. Dick and others. I did not become interested in exhibit design until much later in the fifties and sixties when I enjoyed it because a lot of foreign travel and logistic problems were involved.

In 1935 my biggest project was the design of the exterior of the new Model 100 National Cash Register. This was National's first experience with an industrial designer. The new design with its plain exterior and radiused edges represented a radical departure from anything the company had done before. By this time the Teague office included a model shop and I decided to present the concept in the form of a model or mock-up rather than a rendering. I chose a metallic silver gray color for the case which became the standard color for the entire National line. The Model 100 went into production in 1938 following my model very closely except that the production version was somewhat narrower and my art deco name had been changed to the standard logo.

Walter Dorwin Teague Associates' design for a standardized service station for Texaco.
Dorwin's design for National Cash Register, which would be manufactured for over forty years.

Company records are vague on how many Model 100 machines were produced altogether. Over 6.6 million units had been built by 1963 when some versions were discontinued. Manufacture continued until 1976 when NCR brought out a "Bi-Centennial" version decorated in red, white and blue stars and stripes, but otherwise unchanged. This represented a time span of over forty years from design submission which, as far as I know, is the longevity record for a product by an industrial designer. Total sales were well into the billions of dollars. Many Model 100s are still in use in mom and pop stores in America and all over the world. The Teague office did several electric models in later years which followed my basic design theme and color. One of my first patents was for the Model 100 which lists my father as co-inventor. During the same year I designed a cooking range for the Floyd-Wells Company on which I also received a design patent. Another project was the exterior design of a small tractor for Caterpillar, which never went into production as far as I know but I enjoyed working on it, especially since a tractor was put at my disposal so I could practice bull-dozing in a field out on Long Island.

Every year until 1939, when the first New York World's Fair opened, the number of employees at WDTA increased, reaching fifty-five by 1938. This was the peak era for industrial designers in America. The Great Depression had not been a factor. Many companies decided that, even if sales were down, this was a good time to think about updating their products. New industrial design offices opened each year; the original offices, with their experience and reputations, were inundated with work. Industries not only wanted their products redesigned, they were anxious to increase sales with attractive showrooms and exhibits as well. Every industrial design office was heavily engaged in these activities, with the Teague office probably the most active. From 1933 until 1939 Teague assignments included an Atlantic City display for Bryant Heater, a Chicago Fair exhibit for A. B. Dick, a display room for Eastman Kodak, and for Ford Motor a number of tasks including the Dearborn Museum interior, the San Diego International Fair exhibit, automobile shows in New York, Chicago, Detroit and San Francisco, an exhibit in the San Francisco World's Fair plus the huge Ford building and exhibit in the New York World's Fair. Other major buildings we designed for the 1939 New York World's Fair were Eastman Kodak, U.S. Steel, DuPont and National Cash Register. The interior of the Federal Building at the Fair was ours plus big exhibits for Consolidated Edison and A. B. Dick. Listing all the displays during this period would be tedious but altogether it represented an incredible amount of work.

For example, the interior of the Federal Building at the 1939 Fair was a smaller task since it didn't include the design and architecture of a building. However, most of the furniture was especially designed in the Teague office including the massive table for the dining room which seated about fifty. All the dinnerware, cups, saucers, plates and serving platters included the U.S. seal and gold stars around the periphery. Seven kinds of crystal wine glasses and goblets were designed and specially made by the Libby Glass Company—100 pieces of each. A special piano was also designed by our office and built by Steinway which I believe is still in use in the White House.

For the five major buildings we had several architects on the staff. Most of the buildings, in addition to the exhibits, had executive lounges and dining rooms with special furniture and accessories. I remember the Ford V.I.P. lounge in particular. My father designed the ashtrays which were made in crystal by Steuben. Very simple rectangles about four inches high and twelve inches square with a spherical depression scooped out of the center, they were quite beautiful and were purposely made very heavy to discourage pilferage. The lounge was used only by top Ford executives and V.I.P. visitors but within three weeks after the Fair opened every ashtray had disappeared.

Because there was still plenty of product design work, I spent very little time on the exhibits. There were a few exceptions. In the Ford building the centerpiece was a huge, conical, slowly revolving turntable on which many flood-lit parts of the cars and trucks were placed. It was around 100 feet in diameter and weighed about 100 tons. I was brought in on the discussion of the best types of bearings to support the weight while minimizing friction so an enormous drive motor wouldn't be required. I did a little calculation and announced that, in my opinion, the best way was to float it. At first this was greeted with skepticism and ridicule—an enormous amount of water would be required, etc. However a review of my rough design and calculations showed it could be done with a six-foot-wide annular trench inside the periphery in which a series of steel pontoons floated at a depth of two feet. Depending on how closely the pontoons fitted the trench, only a few hundred gallons of water were needed. A rigid central bearing served as the pivot and a rather small electric gear motor drove it at around one or two rpm. This arrangement worked perfectly for the two years of the Fair.

How is it that I am usually able to come up with solutions like this? I think I have a knack for rapid assessment of the basic objectives or requirements of a product or system, for rapidly marshalling all the possible solutions, and for rating each solution against the objectives. Later on I formalized this process. The objectives were always written down, preferably in a way which didn't restrict the solution, and when we got to the solution we didn't rule out anything, no matter how wild or far-fetched it sounded in the beginning. An important rule was that everyone on the design team had to resist the temptation to sketch ideas or otherwise get locked into any specific solution until we had a chance to stack all the solutions up against the objectives. I was going through this same process in the early thirties but I was doing it mentally which required a certain amount of discipline on my part to avoid premature judgments. In the case of the turntable, the objectives included such items as: the starting and running friction shall be minimum; the support and centering means shall be quiet; the drive means shall require minimum input power; the support means shall be smooth and level to avoid racking of the turntable framework; the support means and the centering means may be separate or combined; the cost of the support, centering and drive means shall be minimum; minimum maintenance shall be required for the two years of the Fair, etc.

Possible solutions included: a) a central set of large thrust and radial bearings to support the entire turntable at the center; b) a series of wheels (truck wheels and tires?) near the periphery, running on a concrete or metal track with the pivot bearing at the center; c) a large circular float in a large dishpan shape; and, d) an annular pontoon or series of pontoons floating in an annular trench with the pivot in the center. Solution a) called for a very deep and expensive foundation which would be especially tricky in the Flushing Meadows where the surface is mud to a depth of about fifty feet. Solution b) was less of a foundation problem but to make a track which would stay completely even and level for two years is not too easy. Solution c) required a lot of excavation and the circular boat would require a lot of ribbing and structure. Solution d) was obviously the best. It didn't require much excavation, was by definition completely level at all times as well as silent and smooth, and the only friction was the skin friction drag which was easy to compute. The "boats" were in separate sections, easy to build, transport and handle. A concrete trench six feet wide and three feet deep was easy to pour and the annular "boats" could be welded steel covered with a suitable primer and anti-fouling paint. My suggestion for the drive was to provide three or four electric "outboard motors" in wells between the boats or pontoons, and since the speed was about three knots a total of ten or twelve horsepower should be adequate. Actually, this drive idea was a bit too much for the construction engineers who settled for a 20 hp electric motor at the central pivot with a large gear reduction.

Another of our product design clients, Powel Crosley of the Crosley Corporation, handled all his jobs personally. We used to see a lot of him in the office. I did some table radios for him and a machine called an Xervac which was supposed to promote hair growth. A metal helmet with a close fitting sponge rubber band inside, it was connected by a rubber hose to a machine which applied alternating pressure and vacuum in fifteen-second cycles. I had no particular hair problem at that time so I can't vouch for its efficacy. One day my father and I were having lunch with Crosley and he told us of an episode which illustrates the problems of the rich. Before leaving on a European trip, he called in his architect to design and build a little guest house on the grounds of his Florida estate while he was away. Crosley had made a sketch of the ground plan and elevation which he left with the architect. The architect rushed the design and construction and, on Crosley's return, proudly took him over to see the finished building. It was faithfully executed to the plan but Crosley had neglected to provide any scale or dimensions on the sketch and the new building was at least twice as big as what he had in mind. Instead of a Petit Trianon he had a veritable Versailles. In later years, Crosley produced a very small car but we never had a chance to work on it.

A minor task I did for Ford was in connection with the design of its main New York showroom at 1710 Broadway. New car models were to be displayed on low platforms scattered around the room which required a railing to keep visitors back away from the cars. This looked like a fun job so I designed the stanchions and the railing. A pattern was made in our shop and the required number of stanchions were cast in bronze, machined, sanded,

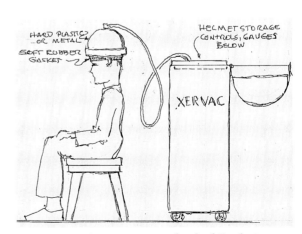

Dorwin's contemporary sketch of Crosley's mid-thirties effort to promote hair growth.

buffed and bright chrome-plated. Delicate tapered elliptical sections, they were singled out for special merit by my father's arch enemy, Lewis Mumford, in his *New Yorker* review of the interior.

Other products I designed were a line of air conditioning units for York and some horizontal internal grinders and boring machines for the Heald Company in Worcester, Massachusetts. We were able to clean these up considerably and make the controls more accessible, as the before-and-after illustrations show. My father was on the Board of Design of the New York World's Fair of 1939 and, just before it opened, the Fair authorities decided that some additional form of transportation was needed. Very little time remained so it was decided to use a small gasoline-engined tractor pulling a string of four little cars. It was my job to design these. I sketched up a little wheeled tractor and some cars with plywood sides and seats and striped awnings. Chrysler Corporation, I believe, built these little trains. My drawings were closely followed. And the trains made it to the Fair on time. There were quite a number of them and, oddly enough, they worked out beautifully, supplying safe, reliable transportation for the two years of the Fair. Despite constant use they were in such good condition after the Fair closed that they were turned over to the Bronx Zoo for people movers. Again despite heavy usage, the Zoo continued to operate them until 1978 when the axles started breaking and the trains were finally retired after forty years of rugged service.

The thrill and satisfaction of working out solutions to design problems continued to be sufficient reward to keep my drinking under control most of the time. While periods of alcoholic behavior still occurred, they didn't affect my work. A mitigating factor was an excellent physical condition due to outdoor sports and exercise.

When I started working full time for my father in 1934, Rita and I had moved down to an apartment on Austin Street, across the tracks from Forest Hills Gardens. My first son, Walter, was born there in 1936, my second son, Lewis, in March of 1938. I commuted to the office at 210 Madison Avenue on

Dorwin's "minor task" for Ford, the people mover for the Fair, "before and after" for Heald.

the Long Island Railroad and I was working hard but still managed to find time for sports and recreation. My brother Lewis had been introduced to skiing at the Gow School up near Buffalo and the sport was just beginning to become popular in America. Hannes Schneider had introduced the Christiana, or stem turn, so for the first time the boots could be held down firmly on the skis and skiers could come down slopes and narrow trails under good control. Lewis and I started out on gentle slopes and graduated to the rope tows at Phoenicia, Gore Mountain and North Creek in New York State. Railroads started running ski trains to these places and then to Stowe in Vermont, Intervale in New Hampshire and the Berkshires in Massachusetts. Every Friday night after work Grand Central Station would be a forest of skis as hundreds of skiers converged to catch the trains to ski country. Before long several others in the office were infected with the bug. Almost every weekend from December through March several of us would join forces to drive or take the ski train to the slopes.

A favorite trip for my brother and me was the ski train to Intervale, New Hampshire, which cost $12.00 for a round trip with a lower berth. It left around 6:00 p.m. Friday night and there was invariably a big party on the train before we turned in to our berths as the powder snow sifted in between the cars. Saturday morning we would be met by a car from the Dana Place in Pinkham Notch. From there we could ski on the Wild Cat trail or on various slopes in the area. Later in the year we would climb to Tuckerman's Ravine, carrying our skis on our shoulders or climbing the fire trail on our skis with seal skins on the bottom which were removed and wrapped around the waist for the skiing down. One spring several of us in the office drove up in late June to the Brick House and went for a swim in Lake Winnepesaukee, then skied in Tuckerman's the next day. The ravine can only be skied safely in late spring after the snow stops avalanching. This vast bowl is a dramatic sight and very challenging up near the top of the headwall where the slope is generally steeper than 50 degrees.

Another favorite spot was Stowe in Vermont where we stayed in a farm house operated by Ma Peterson near the bottom of the toll road. The $5.00 per night included vast breakfasts and dinners plus a box lunch for the mountain. There was no lift at Stowe until 1940 so we climbed the toll road or the Perry Merrill trail and skied down the Nose Dive. Two runs a day was average; occasionally we could do three. In the thirties the Stowe High School team was the equal of most college ski teams. One of their stars, Lionel "Babe" Hayes, was the first person I ever saw who skied parallel. Although only about sixteen, he was a natural teacher and gave us pointers. I finally got so I could parallel ski fairly well.

One of our regular group on ski trips was Ned Scott. Until Head introduced aluminum skis some years later, all skis were wood with steel edges that were installed on order by the shop where purchased. The edges came in foot-long sections and were held in morticed grooves by tiny flat-head screws. Unless they were fitted exactly right, the screws would work loose and the edges would get torn off after a period of use. Scotty installed edges and bindings for Gimbel's and later for Jules André on Vanderbilt Avenue; his edges were the only ones that could be relied on to stay put indefinitely. He was also the authority on bindings which were the old "bear trap" variety (a spring in the heel groove of the boot and toe irons that would not release in torsion, causing many a spiral fracture).

Because he needed a less seasonal job than working in ski shops, I arranged for him to drive one of the cars in the Ford exhibit at the New York World's Fair. Our design for the building included a continuous loop and a spiral ramp around which a stream of Ford and Mercury cars were constantly moving. After the Fair closed in 1940, Scotty worked on the Ford production line in Dearborn and later moved to Sun Valley, Idaho, where he prospered and eventually started to manufacture ski poles.

Road racing had virtually died in the United States until 1936 when the Vanderbilt Cup was revived on an artificial road track laid out on Roosevelt Field on Long Island. This was great news for those of us who followed Grand Prix racing in Europe; for most U.S. racing fans, however, the oval in Indianapolis was the acme. A group of MIT friends and I had entree into the pits at the Vanderbilt where we could examine the fabulous European cars. The race was won by Tazio Nuvolari driving one of the latest Alfa Romeos; Jean-Pierre Wimille in a Type 51 Bugatti came in second, much to my delight as a long-time Bugatti enthusiast. Count Brivio was third in another Alfa.

The Vanderbilt Cup race was held again in 1937. This time complete German teams from Daimler-Benz and Auto-Union joined the Alfa Romeo contingent managed by Enzo Ferrari. Private entries included Alfas, Bugattis and Maseratis. As expected, the laurels went to the Germans, with Bernd Rosemeyer first in the Auto-Union, Englishman Dick Seaman in the team Mercedes second. The surprise third place went to an American, Rex Mays, in an old Alfa Romeo with a U.S. designed centrifugal blower. Heavily subsidized by the Third Reich, the German cars were very advanced in

Dorwin's sketch, entitled "Stowe Style," reflected an affected skiing practice of the late thirties.

Dorwin's photos from the Vanderbilt Cup races at Roosevelt Raceway on Long Island. Above, 1936, with "Wild" Bill Cummins, the '34 Indy 500 winner, on the right. Below, 1937, Auto Union's No. 4, the winning Bernd Rosemeyer car; No. 9, his teammate Ernst von Delius.

design. The Auto Union especially was extremely unorthodox with its 600 hp sixteen-cylinder engine in the rear, huge ventilated four-wheel brakes and full independent suspension. When accelerating through the gears, the sixteen cylinders sounded exactly like someone ripping a sheet. I took some pictures with a borrowed Leica but the best shots were by Ed Marshall with his large old folding Kodak, though the eighteen-inch pointer sticking out in front threatened spectators' hats as he panned along with the cars.

After I had worked steadily for a while, I traded in my '34 Ford for a second-hand SSI four-seat tourer. The SS had a Standard-built engine and Swallow body, hence the name. Rakish in design with typical British cutaway doors, the car was so low I could easily stub out my cigarette on the road while sitting in the driver's seat.

About this time I met Paul Marx, a young Englishman visiting America. He brought with him the first SS Jaguar 100 model in the United States. A very fast, beautifully handling two-seater sports car, it had an aluminum body and fenders and a folding windshield. (For obvious reasons Jaguar would become the company name after World War II.) In 1937 Paul and I both joined ARCA (Automobile Racing Club of America) which would be superseded by the Sports Car Club of America after the war. Other members of ARCA were the Collier brothers (Sam and Miles), George Rand, Lem Ladd and another friend of mine, Lou McMillen, who later became head of The Architects Collaborative in Boston. There were perhaps 200 members altogether in New England, New York and Philadelphia with a varied assortment of cars, including MGs, a Maserati or two, a number of Bugattis, Delahayes, BMWs and modified American cars. (In those days even an MG Midget would instantly collect a crowd in New York

Above: The SSI at the barn in New Hampshire, Dorwin buried in its innards as friend Chet Furnald looks on. Below: Paul Marx talking to John Marshall (in car) as Dorwin looks on. Right, from the top: ARCA days—an MG, Miles Collier exiting his MG Special, Paul's SS 100, Dick Wharton's Maserati, the photos from 1938.

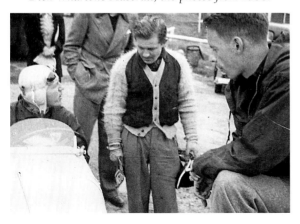

City.) Races were organized on the streets of Bridgehampton (Long Island) and on the Collier estate in Westchester County, as well as rallies and hill climbs up Mount Washington. Paul and I co-drove a rally in the 100 and did quite well.

Pub-crawling in town was one of our favorite activities. Paul was rather small and, with his Cambridge accent, was often picked on in the less fashionable New York bars. Aggressors who carried this too far would regret it since he was the 150-pound amateur boxing champion of England, Ireland, Scotland and Wales, and liked nothing better than an occasional set-to. Paul returned to England as soon as war was declared against Germany. He was killed in one of the early London bombing raids.

I was also a rather steady patron of the Cotton Club and the Savoy Ballroom during the pre-war period. Until the Cotton Club moved downtown to Broadway and 48th, in 1936, both were located in Harlem. It was quite fashionable to visit Harlem spots in the thirties. The Savoy was a long barn-like room with an orchestra platform at each end and tables along the sides. Like the Cotton Club, it drew most of the well-known jazz bands including Fletcher Henderson, Chick Webb, King Oliver, Cab Calloway, the Duke, Benny Goodman and Louis Armstrong. The usual program called for a battle of music in which two orchestras alternated numbers with the winner determined by applause. This naturally resulted in some fantastic jazz. The famous song "Stompin' at the Savoy" was written about the ballroom. The dancing was just as fantastic as the music; the Lindy Hop, Peckin', Truckin' and the Susie Q were all developed there. Very few visitors from downtown assayed the dance floor at the Savoy. The level of dancing was so high and acrobatic that only the local experts could really compete.

The Cotton Club was more famous and somewhat more expensive. Duke Ellington really got his New York start there in 1927; Cab Calloway also made his reputation with "Minnie the Moocher," "Smokey Joe" and "Nobody's Sweetheart." While still a Brooklyn high school student, Lena Horne was in the chorus at the Cotton Club; her father got beaten up when he tried to get her released to join Noble Sissle's orchestra as a vocalist. The movie about the club emphasize the gangster aspect but during my visits there was never any evidence of this, either in Harlem or downtown.

In 1938 my marriage was breaking up and I moved to an apartment in Kew Gardens. One of my co-tenants was Grenville "Gee" Braman who, with Jim Johnson and Freddie Fisher, had organized the Braman-Johnson Flying Service in Hangar B at Roosevelt Field which had been the starting point for Lindbergh and other transatlantic flights and is now a huge shopping center. The field was used by many famous flyers, like Al Williams who used to

make his famous upside-down landing approach, rolling the plane upright just before he touched down. Braman-Johnson's stable included a Curtiss Fledgling trainer and a Ryan ST monoplane used for stunting and advanced training.

Jim Johnson was the instructor, Gee Braman was the business manager, and Freddie Fisher put up some of the original investment. I was not too flush financially at the time but Jim and Gee would give me an occasional free lesson so I spent a lot of spare time out at Hangar B. I remember one Turkish student who was a very nervous type and, after many hours of instruction, was still not ready for a license. Jim and Gee finally decided it was now or never so they notified the local inspector, a man named Harwood, to come over and check him out. Harwood drove to the field in an old Cadillac that was his pride and joy. When it came time for the Turkish student to take his test, the Fledgling was headed in toward the hangar. After Harwood swung the prop to get the engine started, he grabbed the end of one wing to help turn the plane around, since the Fledgling had no brakes. By this time the student was so nervous he hit the throttle but forgot to straighten out the rudder so, instead of a 180, he executed a 360 degree turn and ran into the side of Harwood's prime Cadillac which he had parked nearby. The kid was too frozen in fear to switch off the engine and pieces of running board and bodywork flew through the air as the big aluminum propeller chewed its way into the car. Harwood had to climb up into the cockpit and shut off the engine. Needless to day, the Turkish student didn't get his license.

After the grand total of two hours and twenty minutes of instruction, Jim decided I was ready to solo. Having never taxied the Fledgling, which was somewhat tricky, I was ground looping all the way back to the start of the runway. Though I made some more hops, I was low in funds and I wasn't terribly interested in flying anyway so never went for a license. One summer day Jim, Ed Marshall and I chipped in to rent a three-place Stearman to fly out to Point Democrat on the western end of Fire Island for a swim. There was a certain amount of ground fog on the way out. Ed and I were in the forward two-place cockpit when the top of a radio tower suddenly loomed up in front of us. I turned and frantically pointed upward; Jim, in the rear cockpit, horsed it up just in time. We landed on the beach after dragging it to make sure there was no large driftwood, took off our clothes and swam in the surf. I have a picture of me swinging the prop with nothing on but a helmet and goggles. It will not appear in this book.

Left: Dorwin's sketch of a W.O.-era Bentley in action at Le Mans. Above: Dorwin and a Ryan ST during his brief flying period. His "Student Landing" sketch at top is circa 1937.

Chapter Six

1939 – 1941

I also learned to trust my instinct and to question sacred dogma ...

Younger writers speak of the later thirties as a primitive era, sort of halfway between the Stone Age and today. At the time we thought of ourselves as very modern, and looking back at how we lived and what we did, I'm not so sure we were entirely wrong. Our planes today are bigger, faster and quieter, we have television, calculators and computers, but in some ways we were more comfortable then. When the New York World's Fair opened in 1939, I was in a privileged position, that's true, but many things about the '39 Fair were better than the 1964 New York World's Fair and later ones. Those who could afford it, and it wasn't all that expensive, could take a date to the River Club in Manhattan and be picked up by a fast hydrofoil boat to carry them smoothly to a landing close to the Fair entrance. There they stepped into a Fair train for a free ride to any destination. I took this route a number of times and it was a delightful way to go.

Actually transportation within a fair had been even better far earlier. As I wrote in the *Architectural Beacon* in 1964: "Since 1933 Walter Dorwin Teague Associates has had exhibit buildings in nineteen major fairs and international expositions in various countries including Italy, Holland, Spain, Austria, Yugoslavia, Switzerland and the Caribbean as well as the United States. The biggest single unsolved problem with the 1964-1965 New York World's Fair remains the same as it was in the eighteen others—sore feet. One might even say that the world has gone backwards in this respect; the best solution ever worked out for visitor circulation was in the Paris World's Fair of 1900. In addition to such engineering feats as the Eiffel Tower and the huge ferris wheel (still to be seen in the Vienna fairgrounds), the Paris 1900 Exposition boasted a three-level moving sidewalk, two and a half miles long, which meandered along the banks of the Seine and past all the major pavilions. A stationary lane, a slow speed lane, and a high speed lane provided easy access to all points. Steam was, of course, the motive power and history indicates that the system was safe and reliable. This was without a question the most intelligent approach to crowd control ever worked out, before or since."

At the '39 Fair the favorite eating place was the Terrace Club where my father, as one of the Fair board, had a membership. Grover Whalen was usually there for lunch with some V.I.P. guest; Mayor LaGuardia often ate there as well. In number and size, the exhibit pavilions in 1939 have never been equaled. By general consensus the most popular was the General Motors Futurama which followed an ideal exhibit format. The only hard part was getting in; that accomplished, the spectator sat in a soft upholstered moving chair, with muted high fidelity speakers located in the chair's wings at ear level. The engineering of the transportation system, as well as the sound and lighting, was impeccable. A huge model of the imaginary cities and towns of the future was spread out below the viewer as he drifted quietly and smoothly over the landscape, a well-modulated recording explaining the various features below. The spectator had no decisions to make, he neither had to stand or walk; all he had to do was to sit back and relax while the exhibit did the rest.

This was Norman Bel Geddes' concept but much of the credit for its success lay in the expert manner in which all the details were carried out. And was due in large measure to the work of Roger Nowland, the consulting engineer retained by Bel Geddes. Roger collaborated in developing imaginative but plausible concepts of the future for housing, vehicles, highways and buildings, and he was responsible for the authentic detail with which the thousands of models and their animation were carried out. Roger joined WDTA after the design and construction of the Futurama was completed in 1938 and he worked with me on the Heald machine tool account. It was his idea to carry out the same ogive curve I used on the work head of the grinder in the other machines so as to emphasize the family resemblance. We got along well as we both thought alike about the future of technology and design.

I used to drive to the Fair in my SSI tourer on occasion as it was only a short distance from my apartment in Kew Gardens. The reader might think that maintaining a foreign car was something of a chore at the time but it wasn't because we had Zumbach's on West 54th Street in Manhattan. A Swiss emigre, Charles Zumbach was an expert machinist and had such equipment as internal grinders and Swiss jigborers so he could rebuild or build complete engines. Among his pre-war projects was the bottom-up rebuilding of a 24-cylinder

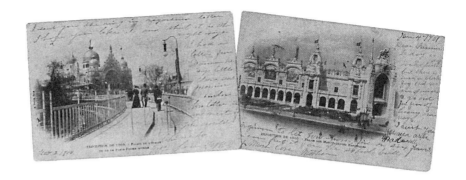

The three-level moving sidewalk from the 1900 Paris Exposition, the Palais des Manufacturers Nationales, the South African building, and the Army Palace entrance. The postcards had been purchased in 1900 by Harriette's grandmother.

Visiting the New York World's Fair in 1939, which in size and number of pavilions was the largest ever. The Terrace Club was the place to eat. Pictured before heading in to dine, from the left, Walter Dorwin Teague, Grover Whalen (the official "greeter" for New York City during this era), Fiorello La Guardia (New York mayor since '34) and Edsel Ford.

Miller marine engine designed to take the world's unlimited hydroplane record. It had never behaved right until Zumbach got it; after Zumbach it ran beautifully. The head of Zumbach's repair department, Jacques Schaerly, was a genius on all makes of foreign cars and had several expert mechanics working for him. The shop was always full of the latest Bugattis, Lagondas, Delahayes and Mercedes. Some of these cars required vast expertise, such as the 3.3 roller-bearing Bugatti which *Le Patron* designed with a crankshaft that came apart in ten sections so the connecting rods could be one piece.

Zumbach and Schaerly permitted a limited number of sports car and racing enthusiasts to work on their own cars at the shop on Saturdays. Lou McMillen and I used to go there regularly. The atmosphere was somewhat like an exclusive club. One Zumbach customer, McLure Halley, owned a beautiful fire-engine red Type 51 Bugatti Grand Prix car, once featured on the cover of the *Saturday Evening Post*. Halley disliked the extremely pronounced camber of its front wheels, typical of all Bugattis, so he had Zumbach's remove most of it. This involved heating and bending the hollow, polished Bugatti front axle, resulting in a completely new front end alignment. Halley took the car to France one summer and made the usual pilgrimage to the Bugatti factory in Molsheim. While he was being taken on a tour of the plant, Ettore Bugatti, *Le Patron*, returned from lunch and immediately spotted Halley's camberless Type 35 parked outside. *Le Patron* ordered his men to bring the car into the factory forthwith, had the axle and wheels returned to their original configuration, and replaced the car in the parking lot.

One time Schaerly told me of a Jaguar four-seat tourer which had rolled over and killed its owner. The car had been towed into the owner's barn near Philadelphia and was for sale, as is. I made a deal with Schaerly: I'd bring it back to Zumbach's, he'd sell it and we would split the profits. A friend and I loaded my SSI with tools, spare battery, etc., and drove to Philadelphia. In a dark shed we found the car; chickens had been roosting in it for a year or so. The battery was replaced, we added oil and fuel, pried one of the front fenders off the wheel with a crow bar and replaced a damaged wheel with the spare. The car started up readily so I drove it around a nearby field a few times to make sure it didn't pull to one side too much and that the brakes worked. The windshield had been torn off but I brought goggles along. The drive back to Zumbach's was without incident.

Later in 1939 my divorce became final and on December 29th I married Harriette Barnard. We had first met in Forest Hills when I was about eight. Karl Easton warned me that my reign as Run Sheep Run champion might be over as a new girl was moving into the neighborhood who was reputed to be the fastest thing on two legs. Actually, I think I usually hung on to beat her but we became good friends. Harriette was a splendid

Above: Charles Zumbach (left), with Jacques Schaerly, in front of Zumbach's on West 54th Street in Manhattan. Page opposite, above: Dorwin's somber pastel sketch of the Jaguar four-seat tourer, drawn from memory after he got it back to Zumbach's. The car had been rolled, wiping out the windshield, doors and top. One wheel was bent and the fenders were crumpled. The owner's widow had stored the wreck in a chicken coop on their farm. Page opposite, below: The Briarcliff Manor AARC race, June 1935, with McLure Halley (wearing goggles and standing to the rear) while his mechanic tends to his Type 51 Bugatti, which had earlier been "Zumbachized." Photos: Courtesy of the Joel Finn Collection.

Harriette Barnard on the day before she married Dorwin Teague. The sketch of Harriette that Dorwin made on a cocktail napkin. Harriette on the slopes at Ste. Agathe in Canada.

athlete. She took lessons from George Agutter at the West Side Tennis Club for many years. Tennis players who didn't know her could spot the Agutter style. Although she played with Kozuleh and other top pros, she never had sufficient killer instinct to become a champion. At Smith she was captain of the hockey, basketball and tennis teams and, unlike me, always managed to get high grades.

We took a lease in a new garden apartment complex in Douglaston on Long Island. In order to save money, I sold my SSI for a fair price; about the cheapest transportation I could find to replace it was a 1932 twelve-cylinder Lincoln dual cowl phaeton in good condition for $150. People just weren't interested in big open cars then. They are coveted today; that same Lincoln is currently valued at $150,000+. For winter driving I had a special partition made which enclosed the front seat and I installed a Sears Roebuck heater. With the side curtains up and the heater going full blast I could

The Teagues' 1932 V-12 Lincoln Phaeton. Dorwin with Harriette (left) and Norman Updegraf (right).

generally keep the front seat temperature above freezing. A cruising mileage of about 10 mpg wasn't prohibitive with gasoline at twenty-three cents a gallon. With the Lincoln's indestructible heavy gauge fenders, I gave no quarter to New York City cab drivers, who always came out second best.

On our honeymoon Harriette and I first went to the Brick House but soon decided that open fireplaces were no substitute for central heating in New Hampshire winters. At Mont Tremblant in the Canadian Laurentians we had a cozy cabin heated by a coal stove that was stoked up every morning by one of the maids. Back home our apartment in Douglaston was on the edge of a pretty salt marsh and a little screech owl lived in a tree outside our bedroom window.

The WDTA office moved uptown to 444 Madison at 50th Street. My biggest account was a series of products for Montgomery Ward, the most important of them a line of vacuum cleaners including a tank type, a floor model and a hand cleaner. I designed the tank cleaner with a "soft" square cross section which permitted the use of low die-cast runners fastened to the shell in place of the wire runners or wheels previously used on this type of cleaner. The end bells were polished aluminum die-castings; a circular plate with the Ward logo was located on either side. In the case of the floor type and the hand cleaner, I invented a new interior configuration. Heretofore all floor-type cleaners had been designed with a scroll-type diffuser around the impeller, which is the classical configuration for centrifugal blowers and which dictates a high, rather clumsy shape which cannot pass under low beds or dressers. Instead I placed the motor with the shaft running fore and aft and a streamlined motor housing which formed a vaneless diffuser or constantly expanding air passage between the outside of the motor and the main housing which opened into a large circular bag connection at the rear. My experience with centrifugal blowers being very limited in 1939, I was acting largely on instinct in adopting this novel configuration. In retrospect I am amazed that Montgomery Ward and the Apex Electric engineers went along with such an untried concept, since the tooling investment must have been enormous, considering the size and complexity of the die castings involved. Luckily, however, it worked out better than any of us hoped, with a performance that was 85% better than Ward's previous model and 49% better than any competitive cleaner. The absence of vanes or scroll cut-off reduced noise considerably and the low overall height now permitted the cleaner to be used under beds and cabinets with low clearance.

The little hand cleaner used the same wrap-around diffuser configuration but the housing was

Public relations photos taken following the successful completion of the Montgomery Ward vacuum cleaner project suggesting collaboration between father and son and mightily irritating Dorwin since his father was not involved at all.

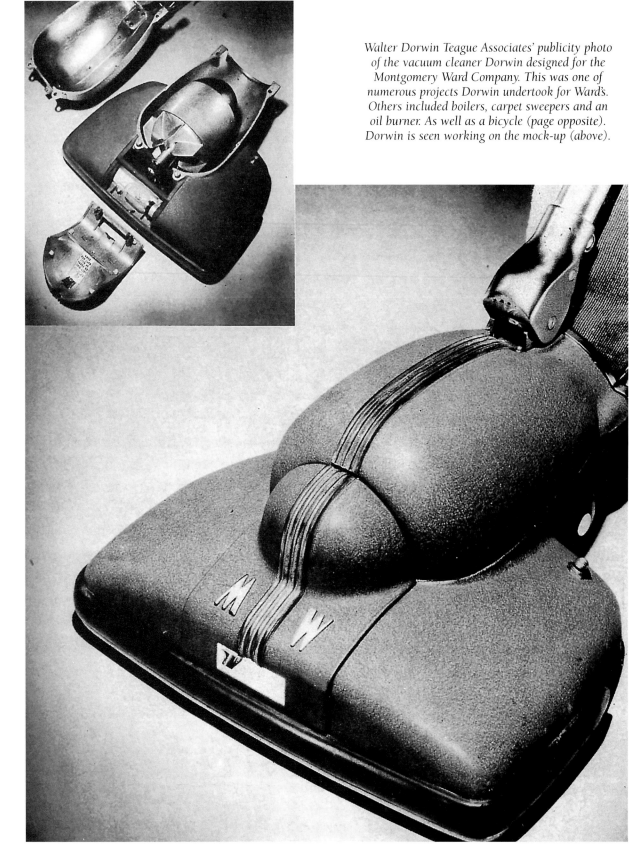

Walter Dorwin Teague Associates' publicity photo of the vacuum cleaner Dorwin designed for the Montgomery Ward Company. This was one of numerous projects Dorwin undertook for Ward's. Others included boilers, carpet sweepers and an oil burner. As well as a bicycle (page opposite). Dorwin is seen working on the mock-up (above).

split on the vertical center line so that the two mirror image die castings made up the entire housing, nozzle, handle, etc. This made a practical and inexpensive arrangement for this competitive market. All three of these vacuum cleaners had excellent performance, were easy to service, inexpensive to manufacture and an outstanding marketing success for many years. The Apex Electric engineers cooperated magnificently throughout this development and the entire line was on the market in a surprisingly short time considering that I started work on the project in 1939. By the end of 1940 sales of the three new models were already 120% above those of the old model for the same period of the previous year. The basic configuration and nearly all of the details of these rather bold and revolutionary concepts followed my original drawings closely. I wrote an article on this development for *Electrical Manufacturing Magazine*, which later was one of the main factors in my being employed by Bendix. (See Appendix C)

How did I know enough to work out these concepts with no prior experience in vacuum cleaners and little training in aerodynamics or centrifugal blowers? I didn't even know about diffusers, or what their function might be. It wasn't

until much later at Bendix that I learned the purpose of the diffuser was to convert the high velocity of the air leaving the impeller into pressure and that I had instinctively designed an efficient vaneless diffuser, the ideal type for a centrifugal blower which must operate over a wide flow range, as a vacuum cleaner does. I guess the answer to that question is that I loved to solve mechanical problems. The satisfaction I received from working out a better solution and finally watching it operate in three dimensions was my greatest reward. I was amazed that someone was willing to pay me for doing something that was so much fun.

For the early industrial designers the satisfaction in taking an ugly product and making it more pleasing aesthetically was the reward. For me the ideal task is one in which I can contribute to the basic function and performance, to making it smaller, lighter, easier to use and better at its job which invariably results in a better appearance as well. One reason I have been able to do this is a knack for determining the ideal objectives of a product, without being unduly influenced by its present form or prior history, and for assigning priorities to these objectives. One of the most important attributes of a floor-type vacuum cleaner, for instance, is its ability to work under low beds, dressers and other furniture. With a vertical motor and impeller having the air inlet at the bottom, excessive height is inevitable. Because of the huge scroll type diffusers used on prior floor-type vacuum cleaners, laying the motor on its side would make it even higher. The height seemed like one of the most important criteria to me; this called for laying the motor on its side, but with a different type of diffuser which would minimize the height. Wrapping the diffuser around the motor in a smooth, uninterrupted, constantly expanding passage not only reduced the overall height but the air flow helped cool the motor, a distinct advantage.

Another factor which affected this and later tasks is the way my aesthetic appreciation has influenced my mechanical solutions. A product that looks complicated is a signal to me to look for a way to make it simpler, which often results in fewer parts and less cost. This fosters a holistic approach to new design problems, consideration of completely new principles rather than merely improving the details of the existing product. I never reject, out of hand, a solution which at first thought appears impossible, if it will provide significant advantages. Too often I've found that "impossible" goals eventually turn out to be feasible, given enough thought and ingenuity. The ability to see both the forest and the trees is something which my aesthetic background has taught me and which is often lacking in people with conventional engineering training. In one of the most useful courses I took at MIT, the instructor pointed out the absurdity of trying for extreme accuracy in computations when the basic assumptions were still vague. He was trying to teach us to take a broad view of the whole problem rather than to waste time perfecting details which might not even be important. This is analogous to the proper approach in painting or sculpture; the artist must have a mental image of the final work of art before spending time on details and must keep the image before him constantly until completion. I learned this at an early age and I'm sure it influenced me to follow the same principle in my development work later on.

I also learned early to trust my instinct and to question sacred dogma, even in the face of opposition and ridicule. In one class in college the professor, in explaining the formulae which generate certain mathematical curves, pointed out the different shapes of parabolas—some long and thin in form, some short and fat. It instantly occurred to me that this was incorrect so I spoke up that actually all parabolas have exactly the same shape. The rest of the class jeered as I stuck to an opinion which I was unable, at first, to put in more precise terms. The professor was indulgent enough to help me define "shape" more precisely, which I finally boiled down to the proposition that for any given proportion of height to width, the angle of the tangent to the parabola at the base will always be the same. He worked this out mathematically and it was correct. The short, fat parabolae are merely enlargements of the tip of the long, skinny ones. This sort of insight is something which only the combined artist/engineer is likely to possess.

Toward the end of 1940 I had begun to get ahead a bit financially so I started looking around for another sports car and finally settled on a second SSI, this time a convertible four-seater. While neither especially fast nor powerful the car handled well and was extremely comfortable, even in the two rear seats. The top folded down to be completely covered by the trunk lid so that the car looked just as well top up or down. Mine was the only example of this convertible version of an SSI that I ever saw, so production of this model must have been very limited. I designed a simple two-piece ski rack which clamped to the left front fender and running board and held four pairs of skis and poles. With my brother and his wife, Harriette and I made many ski trips to the Catskills and Vermont.

The SSI was one of several cars I've owned with right hand drive which I rather like for driving in the U.S. It was actually an advantage for parking and while passing other cars; the principal drawback was that one must pull out further to see if the road is clear ahead. These British four-seater sports cars

Above: Dorwin somewhere in the Catskills with the Teagues' SSI convertible four-seater. Below left: Dorwin and his friend from Stowe, Lionel "Babe" Hayes, with the SSI in New York. Below right: Harriette in the passenger's seat; Dorwin and his brother Lewis flanked by Harriette at Bob Ensign's Catskills home.

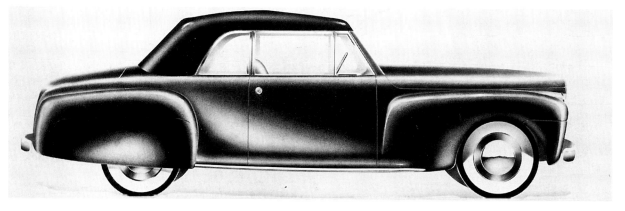

Edsel Ford, ever the aesthete. The Ford that Dorwin designed for him, this version after Edsel decided not to use the existing dies. This rendering, contrary to usual practice in the industry, was drawn exactly to side elevation and is not exaggerated in any way. Some of the competition the Teague-designed Ford would have had in '41: from the top, Packard One Ten, Chevrolet Special DeLuxe, Hudson Commodore.

represent a lost art; no modern sports car provides comparable seating for four adults. The SSI, early Jaguars, Morgans, etc. were low slung and well proportioned but at the same time the two separate rear seats provided adequate head and leg room; two couples could drive 400 miles to the ski country in comfort, providing they didn't bring along too much baggage. It would pay some manufacturer to study these 1938 designs and re-learn how to build a good-looking, compact, low slung sports car to carry two couples on a long trip comfortably.

Fred Black of the Ford Motor Company was the manager of the Ford Pavilion at the World's Fair and was our principal contact during the design and construction phase. Edsel Ford, the company's president at the time, took an active interest in the progress of the Fair building and visited our offices at 444 Madison Avenue a number of times. In 1940 Edsel introduced the new Lincoln Continental which was designed by Bob Gregorie under Edsel's direction. This was one of the outstanding production car designs of all time; no little part of its appearance and success was due to Edsel's influence.

The opposite of his father, who was practical and dogmatic but had no interest whatsoever in aesthetics, Edsel had a warm, generous personality and an infallible aesthetic sense. His style was low key; he made occasional mild suggestions which were usually very much to the point, and he never felt the need to assert himself. He was easy to talk to and, in spite of an obviously heavy schedule as president of the Ford Motor Company, always seemed to have time for discussions of the future of automobile design. During one of his visits to our office, he saw my model from which the Marmon Twelve was derived, which led to discussions about future design directions. Edsel liked my ideas and asked if I would attend some of the national automobile shows and report back to him on my impressions and suggestions. Being paid to attend an automobile show was my idea of a good time. The upshot was a report I prepared for Edsel in which I said the time was ripe for an American sports car and the best way to go about it would be to bring out a special version of the Ford or Mercury, following the lead of the Lincoln Continental cabriolet, only smaller, lower and less expensive—sort of a family sports car. (See Appendix E.)

Edsel agreed and asked me to work up an illustration of what I had in mind. In view of the enormous tooling cost of a completely new design, he suggested that I should employ as many existing body dies as possible. It didn't take long to make some drawings and a rendering which turned out quite well, in spite of the use of current fender and hood dies. When he saw these he reversed his opinion and asked me to carry the idea further forgetting about the body die limitation. My air brush rendering retained much of the Lincoln Continental feeling and simplicity, but was lower and more compact, with fenders more like those I had used in the Marmon Twelve model. Edsel was extremely pleased with the result which was finished around the latter part of 1941. By this time the country was beginning to concentrate on war production so the introduction of a new model wasn't in the picture. Tragically, Edsel Ford died prematurely on May 26th, 1943. His death was a sad blow to the Ford Motor Company and to the entire automobile industry. Unfortunately Edsel is inevitably linked with the miserable car that bore his name. Actually, the Edsel car, conceived a good decade after Edsel Ford's death, was the antithesis of everything he stood for, and would certainly have been prevented if he had still been alive and president of Ford. Approved by Ford president Ernie Breech, the Edsel was a prime example of what happens when a product is designed by a marketing department. Postwar it would also be downhill for the Lincoln Continental which was increasingly gussied up with extraneous ornamentation.

By 1941 it was becoming evident that the United States was going to be involved in World War II. WDTA was starting to get an occasional government assignment. One task I worked on was to develop some thoughts on a global strategy room for President Roosevelt. After setting up a series of objectives I concluded that the ideal centerpiece for such a room would be a huge globe, large enough so that an aircraft carrier to correct scale could actually be seen with the naked eye. I finally settled on a 100-foot diameter which results in a scale of about 420,000 to 1 at which scale a 1,500-foot vessel would be a little less than 1/32 inch long. The biggest challenge was to arrange for close-up or even microscopic examination of all areas. One could scarcely ask the President to lie on his back to examine the situation down in the Antarctic, so the only alternative was to rotate the globe until the area of interest was opposite the viewer, who would be seated in a position halfway up the circumference. This brought up the problem of how to rotate, or even support, a 100-foot sphere in any plane, aggravated by the fact that it would preferably have some three-dimensional treatment such as mountain ranges half an inch high.

The technique I finally decided on was to rest the globe on a 50-foot diameter sponge rubber-covered ring. To change scenes or bring an area up where it could be worked on, air would be introduced into the base of the ring at a rate sufficient to lift the globe until there was a gap of a couple of inches between the globe and the ring, at which time the globe could be rotated manually until the proper area was opposite the viewing point. In order to test this idea I bought the largest globe I could find, made a wooden support ring, and hooked up a vacuum cleaner outlet to the base of the ring with a means of modulating the air flow. Tests showed that the globe always centered itself over the ring and could easily be rotated as desired. If, for example, the globe weighed a ton, the pressure required would only be about .01 psi which would require a relatively small blower.

This scheme, I felt, would supply the President and his Joint Chiefs of Staff with a close-up of any corner of the world in a manner that would bring out its relationship to the global strategy, as no other technique could. All sorts of refinements suggest themselves in a modern version of this scheme, such as computer controlled air jets to rotate the globe to any desired orientation and a "zoom" control so that viewers could greatly magnify any specific scene to obtain more detail. Teams of support personnel would be constantly updating each area as the situation developed. I doubt if this concept was ever seriously considered (it was shrouded in secrecy at the time) but with the various computer aids and visual techniques available today, it might be a useful and viable tool.

Chapter Seven

1942 – 1944

The experienced engineer would be too smart to even consider this kind of solution ...

As 1941 drew to a close I decided I should be working on something more useful for the war effort than World Fairs and vacuum cleaners. Gee Braman and I discussed the situation and agreed that we would start by checking with Grumman on Long Island. We were both offered jobs but, in my case, the salary would have been about a quarter of what I was making so I decided against it. Gee took the job and eventually did very well when Grumman began turning out more planes per day than any factory in the world. The second place I tried was the Eclipse-Pioneer Division of Bendix Aviation in Teterboro, New Jersey. I was interviewed by Dave Gregg, head of the so-called Research Engineering Department at Teterboro, who was kind enough to overlook my lack of any degree. He was especially intrigued by my article on the vacuum cleaners for Montgomery Ward, since his department was currently working on a large centrifugal supercharger for pressurizing the cabins of the Boeing B17s so they could fly at higher altitudes. In April of 1942 I accepted the job as a junior engineer in Research Engineering at a salary about half of what I had been making at WDTA.

Because of its temperament, keeping the SSI running for the duration worried me. In my search of cheap, reliable transportation I came across a 1929 Rolls-Royce 20/25—the company's "smaller" model—that a friend of mine, Hugh Weidinger, had for sale in his showroom in Hempstead, Long Island. It had a beautiful yellow and black lacquer finish on its aluminum body. A very sporting looking close-coupled sedan, the car's aluminum body was finished in a beautiful yellow and black lacquer. Its first owner, William Guggenheim, had kept it in pristine condition throughout. Six brand new tires, a new Cathanode battery, like-new upholstery, a set of crystal cosmetic holders with gold-plated stoppers and the Guggenheim monogram by the rear seat—a nice package even at Hugh's price of $900.00. That was more than I could afford so I offered him half that which he accepted immediately since the car had been in his showroom for several months with no takers. The Rolls turned out to be the most economical and reliable car I ever owned.

Commuting from Long Island with its horrendous morning and evening traffic would have been impractical, so a second problem was finding a place to live, nearer to the Bendix plant. Harriette found an ad in the Sunday *Times* for a house to rent on an estate in Alpine, New Jersey and drove over Monday morning with the agent to check it out. The number of people who had answered the ad was so great that an Alpine policeman was on Route 9-W directing the traffic. The house had been taken hours earlier. As an afterthought the agent showed her a small attached apartment with a garage and sun porch which we took, subject to the approval of Mrs. Rionda. The estate, Glen Goin, was owned by Manolo Rionda, an heir of the Zarnikow Rionda Sugar Corporation. As a sort of hobby, his wife Ellen rented out places on the estate to young couples that she liked. Fixing up some of the servants' cottages at first, she later built about a dozen new stone houses. She also put in a beautiful Olympic-size swimming pool with stone dressing rooms in a secluded wooded area and a couple of excellent "*en tout cas*" tennis courts with a little club house nearby. All the tenants automatically joined the Glen Goin Club which included unlimited use of the pool and tennis courts for just $35.00 per year dues. Later, a couple of paddle tennis courts were added for winter use. The other Glen Goin tenants were young couples our age and there were lots of community parties and activities.

We moved to Glen Goin in April of 1942. Our rent was $45.00 per month until we relocated the following year into one of the stone houses which had two bedrooms, a cathedral living room, dining room and garage with a large storage attic, where we would live for the next twenty-two years, the rent in the meantime soaring to $120.00 per month.

The Riondas lived in a large, gloomy mansion filled with dark Spanish antiques, on the edge of the Palisades across 9-W from the tenants' houses, until the Palisades Interstate Parkway was put through and the house was condemned. Then they built a large new one on our side of the road. Glen Goin had two entrances with the houses scattered in the woods so that each one had privacy. It certainly represented one of the world's great housing bargains but there was one drawback: Mrs.

Lou and Peggy McMillen's BMW 328 alongside the Teagues' SSI on a ski trip in early 1942. The '29 Rolls-Royce 20/25 close-coupled sedan which replaced the SSI later in '42 after Dorwin joined Bendix, again with the McMillens for a ski trip. Note the four pair of skis on the Rolls' fender. Below: Dorwin and Harriette Teague's new home on the Glen Goin estate in Alpine, New Jersey.

Rionda. Forceful and autocratic, she was convinced that her prerogatives included the right to control the intimate details of her tenants' lives. I never had any serious run-ins with her; and I used to kid her a bit which I think she rather liked. Some of the others would occasionally have more serious arguments, however, even extending to banishment in a couple of cases. No one had any influence over Mrs. Rionda except her husband Manolo, a delightful, pleasant person who was the court of last appeal in the case of real trouble. For a while Manolo kept a large cruiser with a captain and crewman in the Alpine Boat Basin at the foot of the cliffs below their house in which he commuted to Wall Street in good weather. Several of the tenants who worked in downtown New York were invited along and served coffee and toast on the way in, martinis on the way home.

During my twenty-five years in Glen Goin, I arranged for several friends to move in. Lou and Peggy McMillen took a house until Lou joined the Navy in 1943. Another was young Jim Campbell who worked at Bendix and his pretty wife, Margot. Jim's father had bought a new Duesenberg—a dual cowl phaeton—but never drove it much and finally gave it to Jim. The odometer read only about 12,000 miles. When Jim left for Europe he asked me to exercise the car occasionally so I would drive it to Bendix every few weeks. After the war Jim and Margot decided to move to California and, of all things, they traded the Duesenberg in on an M.G. Midget. A few miles from Los Angeles they had a head-on collision and were killed instantly. Others I got into Glen Goin were Nicholas Balfour and Ben Stansbury who both worked at WDTA and sailed with me in later years. Ben left WDTA to start his own business in Los Angeles and also got into politics. He was mayor of Beverly Hills for a couple of terms.

Another Glen Goin neighbor was John Halsted who was in charge of purchasing for Colgate and an enthusiastic racing sailor. He owned one of the Herreschoff "S" boats, a 28-foot sloop with a rather roomy cuddy cabin which he raced every weekend at the Larchmont Yacht Club. Like most of Herreschoff's designs, the "S" was fast and seaworthy under all conditions. I became one of John's regular five-man crew and advanced to where he would let me skipper "Fidget" in races when he couldn't make it. The "S" class was quite large then, and the competition was fierce. I became enthusiastic about sailing; though the speed through the water was a fraction of a racing outboard, the excitement was just as great. Because John was an experienced and meticulous sailor, I learned to do things the right way.

Working at Bendix as an engineer was a

Dorwin's first assignment for Bendix—troubleshooting the Boeing B17's cabin supercharger.

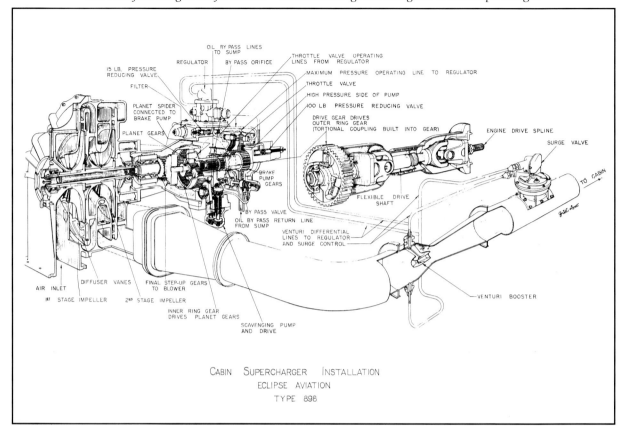

challenge as I was somewhat rusty on the nitty-gritty aspects of engineering which were usually taken care of by the clients in the industrial design business. Simple problems like figuring springs and air flow equations had to be learned or relearned as I went along, but by taking problems home and studying text books, I was able to stay abreast of each task. The fact that this was a new experience made it interesting and the hard work didn't bother me. I always worked best under pressure and enjoyed meeting "impossible" deadlines. The same proclivity made me like racing outboards and sailboats. In business the successful person must often make fast, accurate decisions on a new course of action just as one has to do in any kind of racing. In retrospect it seems I spent a lot of time in extracurricular activities such as sport, nightclubbing and other "frivolous" pursuits. Further I've been a heavy reader all my life, averaging a book a week. But when I look at the volume of work I completed each year at Bendix and WDTA, I am amazed.

One of my first assignments at Bendix was trouble-shooting the cabin supercharger for the Boeing B17s to enable them to operate at higher altitudes where they would be less vulnerable. The large centrifugal blower was driven by a shaft from one of the four engines. Teterboro had a superb altitude chamber where we could dial in pressures down to the equivalent of 50,000 feet and temperatures down to -60 degrees Fahrenheit. Tests of the B17 blower were not up to specification and I was put on the team to get a bit more performance out of it. The Air Force was understandably in a rush so the project had a high priority. Very little was known about centrifugal performance theory then; the best available text was a recent report prepared by General Electric. I latched onto a copy, studied it thoroughly and became convinced that the inlet diameter (about five inches) was too small. From what I know now I don't think my theory was valid, at least for the reasons I gave. No matter, a test unit was built with a larger inlet and the performance came up to spec. Later on the altitude chamber came under my jurisdiction. After the war Washington suggested that if Bendix wanted to keep it, the government should be reimbursed for what it cost. Never at a loss in financial dealings, Bendix said no, we don't want it, come and take it away, just be sure you don't harm our building (which Bendix had paid for) when you remove it. The chamber was an enormous mass about 15 feet in diameter and 30 or 40 feet long, closely encapsulated in its own brick building. Removing the chamber would have meant dismantling the building or cutting up the chamber into impossibly small pieces. Bendix was able to buy the chamber for a few cents on the dollar.

After this I worked on cabin pressure regulators for fighter airplanes which were also being developed during World War II. Bob Flanagan, a fellow member of the Research Engineering Department whose specialty was Rootes blowers, worked with me on pressurizing the Republic Thunderbolt for Paul Pevney at Republic, Bob doing the blower, me the regulator. Our combination worked okay in tests at Bendix but I had to go out to Dayton to demonstrate the regulator in the Wright Field altitude chamber, which was much smaller and cruder than ours. The civilian in charge of the test there implied that if I really had as much faith in the regulator valve as I claimed, I would get in the chamber with it while it was being tested. This put me on the spot. I climbed into the cramped space while the technicians proceeded to bolt the end belt onto the chamber, which took several minutes. Whether it was the size of the chamber or some characteristic of the vacuum pump, the pressure in the chamber which my valve was supposed to be controlling started to surge violently and the altimeter in the chamber was rapidly swinging from below sea level to 30,000 feet and back. Fortunately I had a small screw driver with me which I poked into the adjustment screw on the regulator bellows controlling the pressure curve and this damped the oscillations, after which the valve regulated satisfactorily. Fortunately, my ear drums and sinus passages in those days were able to cope with violent pressure changes unscathed.

The director of engineering at Teterboro was a White Russian named Vladimir Reichel, a cultivated, pleasant, imaginative person. He was impressed by my industrial design background and liked my work. The Germans had a head start on the Allies in several areas including jet propulsion for airplanes, although their jet fighters came along a little too late to make much difference. We were just starting to get into jets and Bendix, traditionally the major supplier of airplane engine starters, was looking at various possibilities for jet engine starter systems. Because the jet engine must be cranked up to a speed of 2000 rpm or more before it becomes self-sustaining, the starting problem is quite different from reciprocating engines and requires much more energy. There were a lot of wild ideas as to how to achieve this kind of power, probably the wildest of all to develop an on-board steam starter. Reichel had been persuaded to give this a try so he handed me the problem. The instigators of this concept were the Besler brothers—Bill and George—who were long-time steam enthusiasts. Among other developments, they had replaced the engine in a Travelaire biplane with a steam engine and made a couple of successful flights. The Beslers convinced Reichel that in order to indoctrinate his engineers in this new element, they should have first-hand experience with steam, the best way to do this being to drive a steam car. The result was that the Beslers leased Bendix their personal Doble steamer and arranged for Bendix to purchase a Stanley Steamer, both of which fell to me as project engineer.

Steam cars had long been gone from the automotive market, the Stanley succumbing in the mid-twenties, Doble (the last survivor) in 1931. I drove the Stanley only once or twice. The starting instructions covered two pages of single-spaced typing; by the time one actually took off, the day was well advanced. The car, about a 1920 model, had a hose built into the right hand running board which could be let down into a nearby brook to replenish the water supply by means of an auxiliary steam pump. The Doble, however, was a different story. It featured a flash boiler, a power burner and working pressure of about 1200 psi. The starting and control system was completely automatic. All one had to do was to turn the key; the burner would rumble and in about a minute and a half the pressure was high enough to get going. The control system was highly complex with a multitude of wires, pipes and solenoid valves and, with the large blower motor for the oil burner, the generator couldn't keep up with the demand for electricity, so I had to put the battery on charge every night. The fuel was normally kerosene but I found that Varsol, used in the plant for cleaning machine tools, worked fine. Every time the tank got low I would go around to the back of the plant and have thirty gallons or so poured in. During gas rationing, this was welcome free mileage.

One of the most attractive features of the Doble was a piercing steam whistle which sounded exactly like that of a steam locomotive. The first time I brought the car home, I pulled up under our kitchen window and blew the whistle which gave Harriette and the neighbors quite a thrill. Visually, the Doble was a very striking looking car, with a long hood and a low two-seater passenger compartment, finished in multiple coats of two-tone maroon lacquer, very much like a Duesenberg in character. In spite of its heavy weight—about 5000 pounds, as I remember—the acceleration was quite impressive and completely silent; the sensation was like being pulled by a giant rubber band. Commuting to work in the Doble was great fun. In my view, the complexity of the steam tubing and control system, and the disaster that would result if the car were to be left outdoors for a

The Doble steamer—a great way to commute to work.

few hours in sub-freezing weather, are the real reasons why steam could not compete with the internal combustion engine.

This susceptibility to freezing was, of course, a prime consideration in a jet airplane which might regularly be exposed to temperatures of minus 60 degrees F. To cope with this, I worked with Ed Smith in our department who was a Fellow of the American Society of Mechanical Engineers and an authority on fluid dynamics and thermodynamics. We finally settled on one of the Freons for which Ed worked the "steam tables" and a Moliere diagram. Freon was not quite as good as water as far as power per pound was concerned but it would have worked. I designed a vane-type motor which packed a lot of power in minimum space and built a tubular flash boiler with water pump and controls. Tests were run with water but, before the entire system could be perfected, the higher-ups decided to abandon the concept for simpler and more reliable methods. The steam starter joined the steam automobile in limbo.

The automatic oil cooler emergency by-pass valve was an urgent project. The Navy was losing planes because the engine oil cooler radiators were quite vulnerable. A single shot or piece of shrapnel would cause complete loss of oil and the engine would quickly seize up. If a valve could be devised that would detect such leakage, it could then act to bypass the cooler on the theory that hot oil was much better than no oil at all. My approach was to use a pair of orifices which measured the flow in and out of the cooler and to compare the pressure drop across each orifice. If all was normal the two pressure drops would agree but if there was a leak, i.e. more oil going into the cooler than coming out, the imbalance would activate a servo valve which would throw the shuttle valve spool and by-pass the

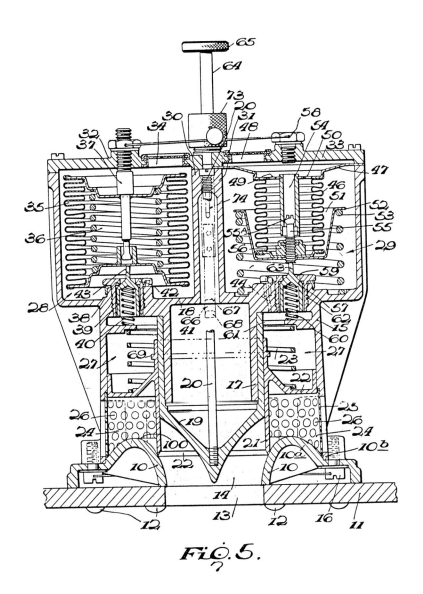

Fig. 5.

oil cooler completely. A prototype was built which worked well in laboratory tests but out in the field on an airplane, it was discovered that negative "G" forces could cause a momentary difference in the pressure drops which would fool the valve into thinking there was a leak and it would by-pass the cooler. To counteract this, I devised a simple negative "G" switch and a reset solenoid so that during negative "Gs" the valve would remain in the normal mode. For testing we mounted one in a Navy SB2C which was rather a clumsy three-place attack plane and for an hour the pilot and I wrung it out trying to upset the valve. This included long slow outside loops to get zero or negative "Gs," inverted flight and hedge-hopping over the Maryland landscape a few feet over the tops of stone walls and trees. This kind of first-hand field experience with our equipment was one of the rewards of working at Bendix, and certainly encouraged dedication and quality control on the part of the engineers.

My records show that I completed about seventy projects during my first three years at Bendix. One device I worked on in 1944 which turned out well was an instrument throttling valve. In those days the turn and bank indicators in planes had gyros that were driven by tiny air turbines connected to the engine intake manifold or to an engine-driven suction pump. Since the suction varied widely a small throttling valve was required in the line to the instruments to maintain the suction at about two inches of water. Because existing regulators were unnecessarily complicated and expensive, a project was initiated to design a simpler and cheaper valve. Existing valves were a variation of the classic regulator which employs a flexible diaphragm to operate a valve stem which throttles the flow of air. A spring on top of the diaphragm regulates the flow to maintain the correct pressure. In this case, the permissible pressure tolerance limits were quite wide so that a calibrated spring would be accurate enough without requiring an adjustment.

I decided to take a completely new direction and started investigating the idea of an in-line valve in which a spring pushed against a sleeve valve which covered or uncovered the holes in a central tube. I reported to one of Dave Gregg's assistants, Don Lawrence, a glider pilot and a good, experienced engineer. Don took rather a dim view of my idea and, to save time, started his own design which was a simplified version of the classic valve. He pointed out, quite rightly, that my scheme would require impossibly accurate concentricity in the machining to prevent the sleeve valve from binding on the inner and outer diameters. I persisted, however, and finally licked the concentricity problem by making the sleeve valve so that it bore on the inner diameter

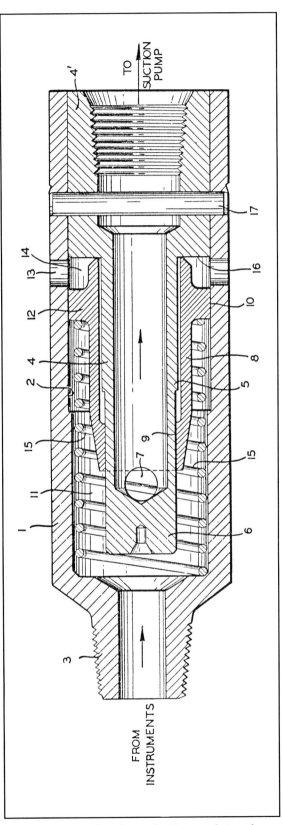

The suction throttling valve Dorwin designed for Bendix, shown about two times actual size.

at one end and on the outer diameter at the other end. If there was some eccentricity the valve would merely cock slightly but it wouldn't bind. The result was a valve with five parts including the pin that holds it together versus some thirteen parts for the simplified conventional configuration. Much less expensive, my valve was also more rugged, weighed about half as much and took up less room. I received a patent on the valve which completely dominated the market for the next twenty years until the instruments went to electrically-driven gyros. The only trouble was that my version was so inexpensive to make, the total profit was not very great. I received thirty-three U.S. patents during the ten years I was with Bendix. Company policy was to pay the inventor $5.00 for each patent, just enough so that he would take the trouble to submit it but not so much that he would neglect more mundane duties in favor of dreaming up far-out ideas.

The little valve was another example of my "artist's approach" to solving a design problem. The conventional type of regulator just didn't look right for the job it had to do. Although fine for maintaining accurate pressures in a laboratory or basement where weight and space are not important, on an airplane it had to be mounted somewhere in the suction tube between the manifold or suction pump and the instruments, in an environment where space is at a premium and weight is even more important. The conventional regulator sticks out quite far to one side, where it occupies space unnecessarily and is vulnerable to catching on rags, tools or clothing when a mechanic is working on adjacent accessories. It is questionable whether a conventional regulator could be trusted to be supported by the tubing itself, so it would require some sort of bracket to bolt it to a bulkhead. These are some of the thoughts that went through my mind as I studied the problem. I always ask myself at the start of any job "what would be the ideal configuration for this gadget if I didn't have to worry about how it works"? In this case the idea would be a cylindrical shape, as short as possible, no larger in diameter than the tube itself or at least no larger than a standard tube fitting. In its final form the valve wasn't quite that small nor was it cylindrical but it wasn't much bigger than a fitting, being made out of 3/4-inch aluminum hex stock to eliminate machining. Overall length was 2-1/2 inches. Extremely rugged, much more so than a conventional regulator, it was light and compact enough to be supported by the line itself without any additional bracing. Unobtrusive and not liable to catch on anything, it was in effect just a wide spot in the line.

The experienced engineer would be too smart even to consider this kind of solution. Not only would he worry about the concentricity problem, he would probably be afraid of sticking or "hunting," the bane of many regulators. The conventional regulator does not leak at all; my regulator allows a small amount of leakage. It would be difficult to incorporate an adjustment (although I did build an adjustable regulator of this type later) and whoever heard of a regulator without an adjustment? The artist/engineer is, in a way, too dumb to know that a thing won't work; he goes ahead and tries it anyway. In my case I never thought of the concentricity

Page opposite: Exhibit at Bendix of various pieces of rotating equipment that Dorwin designed. Left: Speed regulator for the cabin supercharger of the B17 blower, driven by engine, designed by Dorwin. Below: Yet another Teague assignment for Bendix, a motor-driven de-icer unit. During his ten years at Bendix, Dorwin completed approximately seventy projects.

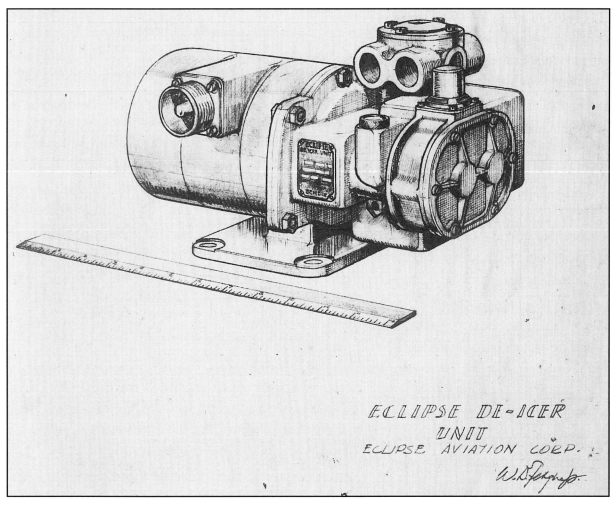

problem and my first design would have required highly expensive machining accuracy. As far as an adjustment was concerned, I knew that the specification was loose enough so that an accuracy of plus or minus 10% was adequate and by paying a nominal premium, springs could be purchased which would regulate well within this envelope. I also knew that a small amount of leakage in this application was unimportant and negligible compared to the flow required to drive the turbine. I didn't expect any instability or hunting tendency since in a sliding valve of this type there is no Bernouili closing effect such as one gets in a poppet type of valve. The only real problem was the concentricity which was solved as described.

The Eclipse Pioneer Division had an impressive engineering and development staff of about 2,500 people. In addition, we had three very complete machine shops, two large drafting rooms and the latest in all kinds of test equipment. Specialists in ferrous metallurgy, non-ferrous metallurgy, chemistry, lubrication, etc. were always available for consultation. There was a refreshing, vigorous spirit about the whole operation; everyone was cooperating and giving their best effort. Bendix's business was automotive and aircraft accessories which covered a vast range of products so that we never had to work in one small area long enough to become bored with it. The general manger of the plant and the director of engineering were both engineers, and the engineers were the top of the hierarchy at Bendix. Some of them took their positions a little too seriously and tended to be somewhat condescending toward the draftsmen and shop personnel. Partly because I was a relative neophyte, I made a point never to disregard or scoff at suggestions from the machinists or draftsmen, most of whom had a lot of practical experience which I lacked. This built up a very useful rapport. Everyone in the shop and drafting rooms learned that I wouldn't laugh at their ideas, some of which were very worthwhile. I usually got fast service in spite of heavy competition and a couple of times their cooperation made the difference between success and failure of my projects.

Other than very rough sketches, few of the engineers made their own drawings. I was the exception; in addition to the necessary calculations I always made some sort of cutaway or layout to scale to show exactly what I had in mind. The engineers turned over their calculations, sketches or drawings to one of several "designers." I had heard of industrial designers, stage designers and dress designers but was unaware that engineering departments had their own mechanical designers. These super draftsmen took the engineers' concepts and turned them into workable drawings. They produced a layout, or detail assembly, of the engineer's brain child, working closely with the engineer as they solved the various problems of materials, finishes, clearances, tolerances, etc. Once the designer had the layout at the point where he and the engineer were satisfied that it was as good as they could make it, detail draftsmen were assigned to make the finished working drawings and detail drawings of the various parts. I mostly used two "designers" at Bendix; Howard Laucks and Maurice Vauclair, both highly competent, intelligent and creative with a natural feeling for good design. Maurice, a French emigré, had vast experience. A competent portrait and landscape artist, he was an expert in gear design, machining practice and all of the nitty-gritty decisions that must be made before a drawing is truly definitive. When I left Bendix to return to WDTA, Maurice followed me and we remained good friends until his death at the age of ninety-one in 1994. Howard was younger but also very talented. Their roles were difficult; they had to be creative but patient enough to go through the long and tedious process of putting the engineers' vague ideas into a definitive form and, if they were successful, the engineer would get all the credit and they would get none. Thus I wasn't too surprised when I learned that Howard had left Bendix after I did to become a vending machine salesman where he made more money with far less effort.

During the war the need for engineers, designers and shop workers became so acute that a number of older, retired people joined the firm. Among them was Finlay Robertson Porter, who must have been at least eighty. The man behind the famous T-head Mercer Raceabout, Porter later manufactured and sold a car called the FRP, his own initials. He was a spry, cheerful member of our department, always ready to lend a hand, no matter how menial the task might be. Naturally I was fascinated by his stories of the infant automobile industry. One of Porter's first designs was a steam car which had the usual two-cylinder piston engine, but he designed his with a 180-degree crankshaft instead the 90-degree arrangement used by everybody else because he felt this would make for a more even, regular impulse and smoother running. A steam car lacked a transmission so the engine was connected directly to the rear wheels. On his first road trial Porter discovered why other manufacturers had gone to the 90-degree crank. When he backed into a small ditch, the engine, by chance, stopped on top dead center and the car was completely immobilized.

Another retiree was a tall, lanky, seat-of-the-pants machinist named Emil Hackenberg. I had designed revolutionary three-stage centrifugal cabin pressure supercharger which was supposed to produce a three-to-one pressure ratio at 35,000 rpm with an

 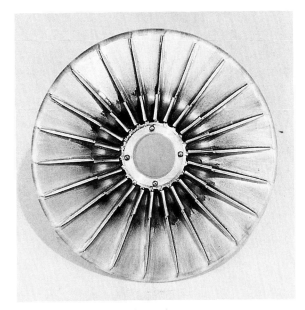

Ram air-driven source of electrical and hydraulic power for high speed missiles or jet aircraft: the diffuser (left), the welded impeller (right), the unit disassembled (below).

air flow of five pounds per minute at 40,000 feet altitude, to give a cabin altitude of around 13,000 feet. The diffusers for each stage were wrapped around the back of the impellers which made for a light, compact unit of this capacity. Carried away by my success with the Montgomery Ward vacuum cleaners, I designed the unit with vaneless diffusers. The shop did its usual meticulous job of building the parts and the unit was ready for testing in record time. To say that the test was disappointing is an understatement; my brainchild hardly put out enough air to blow out a match. This was the most elaborate and costly project I had undertaken so far,

so I was shocked and dismayed. The only possible way to improve the performance that I could think of was to try adding some vanes to the diffusers.

Adding vanes to the parts that were already built was a mean job as there were three different types, which were sort of crescent shaped and had to be fitted somehow to three different conical surfaces. I had no idea how the vanes could be machined or how they could be fastened to the surfaces but I made a hasty sketch of each vane shape and took it to Sam Gilbert, the head of the shop. The only free machinist was Emil Hackenberg whom Sam characterized as good but slower than the "second

Playing hard after working hard ... Dorwin and Harriette dressed up for one of the costume parties at the Glen Goin estate. Page opposite: High jinks at the "cross dressing" party. Mrs. Rionda is in the front row chucking the chin of one reveler, Sir John Templeton is in the center of the rear row. Costume parties were held annually on Labor Day weekend, another event with a "law and order" theme below. Parties were regular weekend activity.

coming of Christ." I gave Emil the sketches without much hope that the fix would be very effective even if he could figure out how to do it, but to my surprise he came to my office a couple of days later with the three finished diffusers, vanes perfectly executed and riveted securely in place, looking like they had been designed that way in the first place. The result was a completely different animal, tests at 40,000 feet showed a compression ratio of 3.3 to one, and an air flow of seven pounds per minute at 34,800 rpm. This experience pointed up the importance of maintaining good relations with all echelons of the company. The sketches I gave to Emil were incomplete and the vanes had to be mounted with very small clearance next to the 35,000 rpm impellers. Brains, experience and a real desire to help me out brought this happy conclusion to the project.

Just about everybody at Eclipse-Pioneer worked hard for the war effort. In our free time we played hard as well. At Glen Goin there were parties every weekend. Down the road a half mile was the Rio Vista estate, owned by Manolo Rionda's uncle, the founder of the Zarnikow Rionda Sugar Corporation and the family fortune. Rio Vista was even more elaborate than Glen Goin with stables, peacock runs, and a high stone tower complete with elevator and billiard room at the top. "Uncle Rionda," as he was known locally, was about ninety when we moved into Glen Goin but he still rode one of his horses every day around the two miles of road on his property. Rio Vista also had a few rental tenants including Scrib and Anita Butler, who lived in the spacious chauffeur's quarters over the original five-car garage.

Scrib was a great fan of Eddie Condon's orchestra. One New Year's Eve he threw a big party for most of the Rio Vista and Glen Goin inhabitants, to which Ralph Sutton, Eddie Condon's piano player, was also invited. Scrib's piano was a decrepit upright and Sutton, who was in no better condition than the rest of us, decided that it needed an overhaul before he would play any more. Accordingly, he spread a sheet on the garage floor and completely disassembled all the keys, hammers and the works which he laid out in orderly rows on the floor. This would have been a monumental job for a sober mechanic under controlled conditions but, incredibly, Sutton put it all back together again and played happily until some time near dawn.

52nd Street was the jazz center of New York in those days with such cabarets and restaurants as Hickory House, Famous Door, 21 and the Onyx Club. Although I visited the Hickory House occasionally to listen to Teagarden and Trumbauer and the girl who played the harp, my favorite was John Kirby's group at the Onyx Club. Kirby was the leader on bass, joined by Charlie Shavers, trumpet; Buster Bailey, clarinet; Russell Procope, alto sax; Billy Kyle, piano; and O'Neill Spencer on drums—"the best small band in America," as was their claim. Kirby kept a tight rein on the group which played subtle arrangements relatively softly with a bouncy rhythm. Billy Kyle had a style quite different from the Duke or Art Tatum and was my favorite jazz pianist. The band played on a raised platform under a low ceiling on which O'Neill Spencer occasionally used the brushes during his terrific drum solos. Russell Procope later took over the number one clarinet spot from Barney Bigard in Duke Ellington's band.

Fabian Bachrach portraits of Dorwin's brother Lewis and his wife Mary Lee Abbott, and his sister Cecily. The 20/25 Rolls-Royce as moving van when the Teagues moved to the stone house.

When the United States finally declared war, Lou McMillen and Ward Bien joined the Navy; my brother Lewis, who had recently married Mary Lee Abbott, a striking model, went into the Air Force, and Dick Crowe, who had married my sister Cecily, became an officer on Eisenhower's staff in London. Dick and Cecily had a beautiful German shepherd named Addie, the most intelligent dog I've ever known, who joined our family for the duration. When I got home from work, Addie brought my slippers from the bedroom, one by one. He got along with everyone at Glen Goin and knew instantly when someone didn't belong there. Once a neighbor called to say Addie had two unsavory looking trespassers up a tree where he was holding them until I showed up. Addie vacationed with us at Brick House where he liked to dive for white stones in the lake. Addie always worried when someone drifted out in the little plywood dinghy I made for my sons T and Lewis who were visiting; he would dive in, swim out, grab the rope in his teeth and tow the boat in to the shore. No one taught him to do this—he just figured it out for himself. One time I took Addie for a walk on a nearby mountain where he strolled off and encountered a porcupine. Since he had never seen one before, he went for it. By the time Addie got back to me, he was a horrid sight with quills bristling from his nose, cheeks and inside his mouth. I ran with him all the way back to my car at the bottom of the mountain and raced him home where I called Uncle Widmer, the Newark surgeon. He in turn called the drug store in Wolfboro and I tore in and back with a can of ether. After Addie was knocked out, we stretched him out on a plank, while Widmer and I went to work with pliers. The grisly task took us at least a half hour before all the quills we could find were yanked out. Poor Addie was a sick dog for a day but recovered completely.

In late 1943 Harriette and I moved out of our gardener's cottage apartment into one of Mrs. Rionda's stone houses. We needed the extra room. Our son Harry was born in August of 1944.

*Top left: Dorwin building a dinghy for sons T and Lewis. Top right: Addie towing the dinghy.
Below: Dorwin and Harriette with Addie; Uncle Widmer, Addie's "personal physician."*

Chapter Eight

1945 – 1947

He saw a big pool of red fuming nitric acid seeping onto the floor ...

As the Germans bombarded England with the V1, and in later 1944 and early '45, the V2 rockets, Bendix began to take an interest in guided missiles and rocketry. The V2 (our designation, the Germans called it the A4) was the world's first long-range ballistic missile, a true rocket using alcohol as fuel and liquid oxygen (LOX) as the oxidizer. This required a pair of large centrifugal pumps, driven by a hydrogen peroxide powered turbine. Guidance was by on-board gyros during the boost phase, which altered course by means of servos that inserted carbon "paddles" into the main rocket exhaust to control roll, pitch and yaw. After the end of the boost phase, the rocket continued on a ballistic curve to its target. In late 1944, we started to get bits and pieces of the V2, classified secret, which were interesting but we didn't know much about the rest of the rocket. We also got parts of an engine for a German rocket-powered airplane that would not become operational before the German surrender.

At that time the Research Department was split into two sections, both under the leadership of Dave Gregg. My section, headed by Don Lawrence, was sort of a catch-all group. The other section, concerned principally with controls, was in the charge of Howard Alexanderson. Among its engineers was Bill Moog. The conservative Alexanderson relentlessly tried to get Bill who was clever and inventive, to forget his far-out notions and stick to well-known principles. Finally Bill, acting more or less contrary to orders, persisted in developing one of his new ideas and was fired. So he started his own company which grew into one of the largest hydraulic control manufacturing businesses in the U.S.

A shake-up in the Bendix organization brought in Ernie Breech, later to be President of Ford, as President of Bendix. Reichel was superseded as Director of Engineering by Roy Sylvander; Dave Gregg was let go as head of Research Engineering. I was never involved or interested in corporate politics so it came as a complete surprise to me to be informed that I was now Senior Research Engineer and head of the Research Engineering Department with a significant increase in salary and responsibility. My heavy drinking periods didn't cease, but again interest in my work kept them under control, and obviously Bendix regarded me as an asset. The department which I now headed was quite small with about fifteen or twenty people including secretaries. I inherited responsibility for the altitude chamber building which was put in the charge of Ed Manning. Two super machinists, Frank Koelmel and Herb Fournier, worked in white outfits at tables in that building and assembled the various gadgets we developed in the department. Frank and Herb were the elite of the Bendix machinists and their expertise and suggestions were a key part of the development process.

The Navy came to Bendix in 1945 for help in developing a new generation of its first rocket, a surface-to-surface missile called the Lark. This was Bendix's first experience in missiles or rocketry and the project was assigned to me. A relatively small bi-propellant guided missile, the Lark used aniline for the fuel and red fuming nitric act as the oxidizer. These propellants are hypergolic; that is, they ignite spontaneously on contact. They were carried in separate tandem tanks and a third tank at the forward end carried compressed air to force the two propellants into the combustion chambers of two rocket motors. One motor had 200 pounds of thrust and ran constantly; the other motor had 400 pounds of thrust and was turned off and on during flight to maintain a constant Mach number. The motors and their propellant valves were supplied by Reaction Motors in Wayne, New Jersey near the Picatinny Arsenal.

The Lark was the brainchild of Robert Hutchings Goddard who was the first liquid propellant rocket pioneer in the United States and actually the world, although the Germans later caught up and quickly surpassed anything done in this country. Goddard had died before we got into the picture. The project was under the direction of Lieutenant Commander Robert Truax, a disciple and assistant of Goddard's who had worked closely with him. Truax was a Navy pilot who flew fighters but was assigned to Lark development because of his rare experience in this new science. One of Goddard's ideas was to lighten the weight of the Lark missile by substituting a turbine-driven pump to provide the necessary 450 psi injection pressure instead of pressurizing the tanks. A significant weight saving would be achieved because not only would the heavy, high pressure air tank be eliminated but the two main fuel and oxidizer tanks could be much lighter since they would no longer need to withstand the 450 psi pressure. The Germans used the same reasoning in the V2 except that they drove the pumps with a third fuel, hydrogen peroxide.

Goddard's concept, which Bob Truax had been working on, was to drive the pumps with a turbine wheel which would be immersed in the jet from the 200-pound thrust rocket engine, the one that was firing continuously during flight. This was a daring concept as the rocket exhaust jet is about as hostile an atmosphere as one can imagine; the noise, even from the little 200-pound motor, is so deafening that ear plugs are required, the jet velocity is Mach 3 (three times the speed of sound) and the stagnation temperature is around 6000 degrees Fahrenheit. No metal will withstand this temperature for any length of time but since the missile's maximum altitude was 20,000 feet, Goddard and Truax figured that turbine blades would only be in the jet for a v short period and the rest of the time they v cooled by the ambient air stream. The diameter the 200-pound motor jet was about two inches and the turbine wheel diameter the Navy used was five inches (we used the same diameter) so only about three blades were in the jet at any one time and hopefully they wouldn't be in there long enough to be burned off.

Altogether this development involved a great deal of fairly sophisticated calculation. First of all a pair of pumps had to be designed for fluids of two very

Propellant valve and pressure control valve for Lark turbo-pump rocket engine.

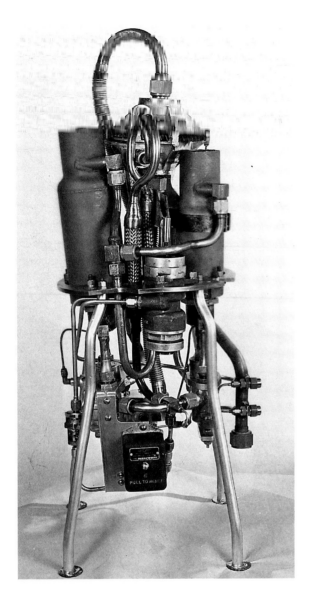 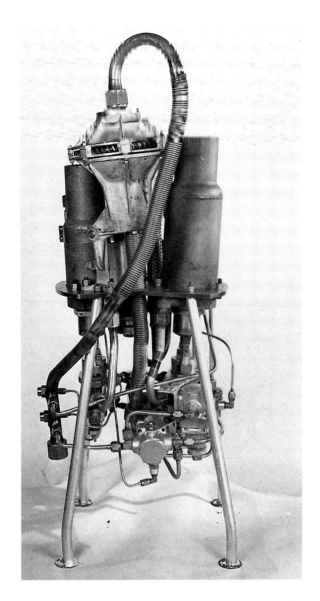

different densities which would produce equal pressures of 450 psi at the correct mixture ratio at two widely different flow conditions equivalent to 200 pounds of thrust or 600 pounds of thrust when both motors were running. One of these pumps had to pump 95% red fuming nitric acid which would be at the bottom of anyone's list of favorite fluids. The turbine presented an even greater challenge. As far as I know, no one has ever tried to run a turbine in a rocket exhaust jet before or since. We were dealing with a technology for which there was no prior experience or guidelines.

The non-technically minded may wish to skip over some of the following paragraphs, but I am still astonished that we came out of this project as well as we did and I feel it is an episode worthy of recording. "Fools rush in …" applies to the fact that I took a fixed price contract for this development for which the pay-off was a successful programmed three-minute run on the static test stand. Obviously I had come a long way in the three years since I joined Bendix as a junior engineer. As leader I devised most of the basic concepts but this wasn't a one-man effort; I had the help of a number of experts in the various fields involved in the development. One of the most rewarding parts of this project for me was the welding of the participants into a group effort, sort of a super brain, like the mass intelligence of an ant colony, which far surpasses the individual intelligence of any of its members. Everyone on the project—secretaries,

Lark rocket motor c.1946. Maurice Vauclair made the incredibly complicated drawings for this revolutionary Robert Goddard concept.

technicians, machinists, draftsmen, engineers and Navy reps—thought alike. The enthusiasm generated in a project like this is best portrayed by Tracy Kidder in his great book, *The Soul of the Machine*, in which he describes how the development staff neglected meals, sleep and their families as they became more and more wrapped up in making the new computer a success. It is the same kind of joy and satisfaction that the crew feels in a long-distance ocean race as they undergo otherwise unacceptable conditions and hardships.

We had Doc Waldman for advice on exotic high temperature alloys and Ken Halliday for coaching us on acid resistant materials of all kinds. Eclipse's fantastic shop people figured out how to machine the thin turbine blades in the almost un-machinable alloys. Maurice Vauclair made layouts and details of the many parts. Laying out the assembled engine was like making an accurate drawing of a plate of spaghetti. Steve Orban followed the tests. Frank Gaffney, my companion and assistant in outboard racing, incurred the wrath of most of the wives of the Reaction Motors personnel who, at Gaffney's urging, worked 'til all hours six or seven days a week. Ed Manning ran the tedious series of pump tests in our lab. Bob Truax contributed much sound advice and encouragement; Reaction Motors supplied the static test stands and helped with the development.

Before we got into the picture the Navy lab headed by Goddard and Truax had actually built some pumps and turbines which never lasted more than a few seconds but provided a good starting point. Bob Truax had used the theory of the momentum created by the deflection of the jet over the turbine blades to calculate the size and shape of the blades. In my completely different approach, I treated each blade as a tiny airfoil and used supersonic airfoil theory to arrive at the shape and size of the blade to achieve the required speed and power for the pumps which I calculated at 29,000 rpm and 18 horsepower to the high flow condition. Airfoil theory is always looking for the best lift to drag ratio which is what we wanted here. My first task on the turbine was to become acquainted with supersonic airfoil theory which was also in its infancy; the best material I could find was a fourteen-page British pamphlet, quite readable and complete enough for practical application. I decided on a symmetrical elliptical section with 30 blades, each 5/8-inch long and 3/8-inch average width, slightly tapered. These represented a nasty machining problem in the very tough, high temperature alloy; finally the spaces between the adjacent blades had to be filled with wax as each new blade was machined. We tried some cast wheels but they didn't stay together and we didn't have time to experiment further.

As for the pumps, this required some frantic juggling of parameters which could more easily be done today on a computer. Each fluid had a quite different specific gravity and the mixture ratio required a volume flow of acid about twice that of aniline. The pressures of each had to be 450 psi when the 200-pound motor was running alone but then, when the 400-pound motor cut in, the flow from each had to triple, still maintaining the right mixture ratio and equal pressures of 450 psi at a higher speed. I calculated the pump dimensions and came out very close to the desired performance on the second try. The turbine and the two pumps were in an aluminum housing with the turbine in the middle and a pump on each side with their inlets on the outside center line. The housing was cut away for the sector where the turbine blades intersected the jet. The bearings for the pump/turbine assembly were the two-impeller inlets, plain bearings lubricated by the propellants themselves. Matching the two pumps took some time and considerable testing but the biggest problem turned out to be the bearing on the acid pump side.

Static firing tests were run at Reaction Motors' new test area at Picatinny Arsenal. RMI was, I believe, the earliest private rocket development organization in the U.S. In 1945 RMI's president was Lovell Lawrence. Other early rocket pioneers there were John Shesta and Jimmie Wyld, who had worked on such stunts as a rocket-powered bicycle and a rocket-powered ice boat in their early days. When I first visited, they were located in an old silver factory near Picatinny Arsenal but were being sued by irate neighbors who disliked the constant red fuming nitric acid fumes and having their walls cracked by the noise and vibration of the static rocket tests. By the static testing stage for our motor, Shasta and Wyld had moved onto Picatinny Arsenal grounds into a series of quonset huts. The quonset huts had a heavy reinforced concrete barrier across the middle with a mirror arrangement whereby the tests could be observed in relative safety. In the event of an explosion, not at all unusual in the early rocket days, the quonset hut would disintegrate easily so as not to confine the blast. At frequent intervals on the grounds, handy open shower stands and eye wash basins stood ready in case of acid accidents.

One major component of the Lark turbo pump engine was the pressure control actuator which constantly adjusted the immersion of the turbine blades in and out of the 200-pound rocket jet to maintain 450 psi outlet pressure on the pumps. The turbo pump assembly was mounted on a hinged bracket and the actuator sensed the aniline pump outlet pressure against an adjustable spring so that, if the pressure exceeded 450 psi, it would withdraw the turbine blades slightly or, if it dropped below 450 psi, it would immerse the blades deeper into the jet. Thanks to my earlier experience with cabin pressure regulator valves, I was able to design an actuator which operated perfectly and never gave any problems with hunting or sticking.

As a result of Maurice Vauclair's unbelievably complicated but accurate layouts, everything went together with no interferences. We were ready for the first static tests. With such complicated systems, one never knows the problems that will be encountered but it is very rare for everything to go as planned. In our case the major offender was the turbo pump bearing on the acid pump side. As one might expect, a fairly heavily loaded plain bearing, running at speeds up to 30,000 rpm and lubricated with 95% red fuming nitric acid, could turn out to be troublesome. Background literature was non-existent. With every bearing material we tried, the bearing seized up after a few seconds of running, the turbine and pumps stopped, and in milli-seconds the 6,000 degree jet carved out a circular notch in the turbine blades as neatly as if it had been done with a grinding wheel. We tried every material imaginable including heavy plating of pure gold and a set of solid synthetic sapphire bearings. Nothing worked.

During these tests Frank Gaffney practically lived at Picatinny and kept the RMI crew working at a pace they had never known before, overtime stretching well into the night. Steve Orban backed him up. Their dedication eventually led to success. I remember one day about noon Steve called me from Picatinny to ask if he could take the afternoon off. I said sure. He explained that while waiting for the RMI crew to finish hooking up the motor to test the latest bearing material he had been sitting with his feet on the desk reading a comic book when he happened to glance down and see a big pool of red fuming nitric acid seeping onto the floor of the control room from a crack under the door. He dove through the flimsy control room window, screen and all, and understandably needed a little time to calm his nerves.

Finally someone suggested Babbitt metal for the bearing which sounded much too easy to be true, after all the exotic materials that had failed. But, oddly enough, it was the answer. We had no more trouble with sticking and finally, barely within our deadline and our budget, we passed the Navy acceptance test. The noise and fury of even these small rocket firings is hard to imagine unless one has actually been witnessed. That anything can stand up to this much concentrated energy seems impossible. We all heaved a sigh of relief when the test was over. I believe a number of turbo pump Larks were built. Some years later, apparently, the Navy declared them surplus; one could actually buy a complete Lark turbo-pump engine in brand-new condition. Bob Truax went on to other projects, got into private industry and finally retired to his native California. The last time I heard from him was after he had designed the rocket that Evil Knievel used for his motorcycle leap across the Grand Canyon.

The Lark project was Bendix's and my first venture into rockets and missiles, and led to a number of on-going projects for the Navy, Air Force and Army in rockets and ram jets. The corporation

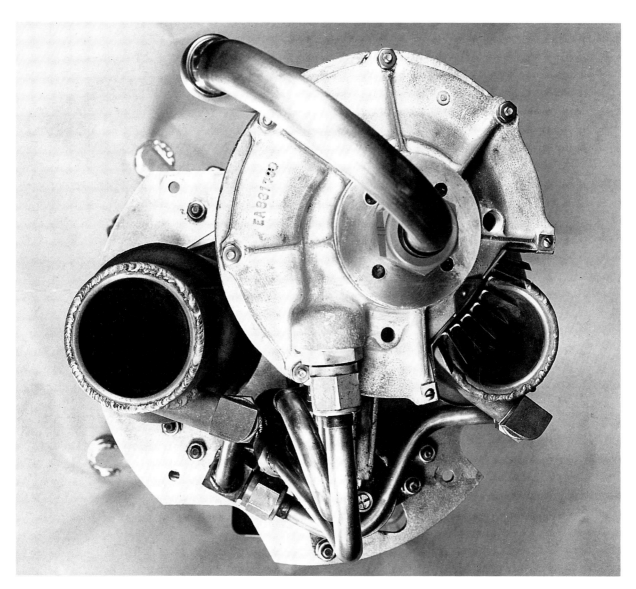

Actual rocket motors and the turbine wheel that was driven by the smaller of the two motors.

Disassembled Lark rocket turbo-driven turbine pump including acid and aniline impellers. Turbine wheel from turbo-pump engine driven by 6000-degree rocket jet at 24,000 rpm.

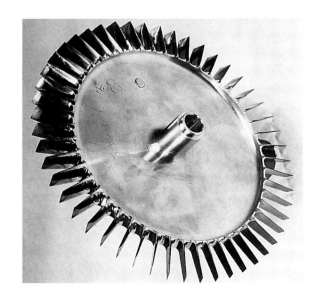

not only got a lot of useful information which could be applied to starters and other products, but we were conducting this extensive R&D at a substantial profit and the number of personnel in the Research Department was expanding.

After V-J Day in August of 1945 our friends started to return from overseas. As expected, the Navy boys brought home the most impressive loot. Lou McMillen returned with a Zundapp motorcycle and side car equipped to carry five Wehrmacht personnel. Ward Bien, who had been port director of Nagoya, had Samurai swords, carbines and tapestries. Brother Lewis, who had fought the war in the Texas desert as a gunnery instructor, had gotten hooked on duck shooting down there and introduced me to the sport. On the edge of a marsh on the Hudson River just north of Snedens Landing we set up a duck blind which we used every fall for

Setting up the duck blind. Below: the victors with their spoils—Dorwin, Lewis and Beany Salembier, all of them holding the same ducks.

several years, much to the dismay of my mother who was practically a charter member of the Audubon Society. Another dedicated hunter was Harry Drinker from Philadelphia, who ran the Pioneer machine shop and had hunting licenses in several states. We used to shoot by the Overpeck Creek in the Hackensack marsh. On good duck days (rain or better yet, a little sleet) we would meet at his house in Englewood at 4:00 a.m. in our duck clothes, drive the Rolls to the nearest point in the marsh, shoot until around eight, change our clothes in the car and arrive for work at Bendix at nine.

Among the many projects we completed in Research was a self-contained power supply to provide emergency electric power for a short period of time in an airplane, if all other sources failed. It had to be capable of indefinite storage with zero maintenance and yet be instantly ready when needed. My solution, for which Al Volk and I received a patent, was a cylindrical compressed air tank with a self-contained pin wheel turbine and a generator with a speed control. Director of Engineering, Roy Sylvander, called me into his office one day to alert me to be ready to show the design to a representative of the Curtiss-Wright Corporation. This infuriated me and I asked Roy why we should show our device to a competitor. Why not let them invent their own? Roy explained that Ernie Breech, still Bendix president, had promised the Curtiss-Wright president that he could have the information on the unit and we had no choice but to show it to their rep, whose name was Villiers and who would be visiting us in a few days. I finally agreed but made a private vow to run it past the Wright man so fast that he wouldn't be able to remember anything about it.

Sure enough, the following week the Curtiss Wright man arrived and turned out to be a tall, thin, very British type with a Cambridge accent and vague "Oh, I say" mannerisms—pleasant and polite, but obviously not very fast on the uptake. Accordingly, I gave him a quick look at the drawings, answered a couple of questions and got him out as soon as possible, with a feeling of a job well done. Almost a week later I got a letter from Villiers with a drawing enclosed that he had prepared and a note asking if I would mind correcting any mistakes he might have made. To make a long story short, he had missed nothing; his drawing was virtually an exact representation of my device, correct in every detail. The dumb Britisher act was just that; he obviously had one of the sharpest minds for engineering and design that I had ever happened upon. I got back to him immediately and asked why he didn't come over and work for a good outfit, which is just what occurred a few weeks later. This was the start of a long business relationship and friendship that lasted until Amherst died in 1992.

Charles Amherst Villiers is another example of the artist/engineer who excelled in both fields. In the 1920s he prepared the Brescia Bugattis for Raymond Mays who was European hill climb champion for many years; Amherst was getting so

The summer of '46, relaxing in New Hampshire, Dorwin, toddler Harry, Lewis and wife Mary Lee. Page opposite: Charles Amherst Villiers with Pope John Paul II in 1982 and the 4-1/2 Litre "Blower" Bentley that he and his son beautifully restored in the early sixties in California. Photographs courtesy of Janie Villiers.

much more power out of the cars that Ettore Bugatti hired him to come over to Molsheim for a few months to help them set up to grind the Villiers-designed camshafts. Later he got into supercharging and designed the units for the famous "Blower Bentleys" that were popularized in Ian Fleming's "James Bond 007" novels. He also worked up the car with which Malcolm Campbell raised the world's land speed record to 206 mph in 1928. Amherst was a first cousin of Winston Churchill and a great grandson of the Duke of Marlborough. He stayed with Bendix until some time after I left in 1952 and then went to Lockheed to take charge of its lunar landing program. Subsequently, for a change of pace, he joined Portraits, Inc. in New York and subsequently moved to Majorca where he spent most of his time painting. One of his last assignments was a portrait of Pope John Paul II.

Chapter Nine

1948 – 1950

This was a useful lesson in how not to build a rocket test stand ...

Late in 1947 the Army approached Bendix with a concept for a new anti-aircraft rocket system, based on a German missile called the Taifun that was just about to become operational at the time of the German surrender. The Taifun had been developed at Peenemunde under the direction of an ex-Luftwaffe pilot and engineer named Klaus Scheufelen. At first Peenemunde was in the hands of American troops although it lay in the Russian zone and was slated to be turned over to the Russians as soon as they arrived there. Thanks to the foresight and initiative of U.S. Colonel Holger N. Toftoy, who was on the spot and couldn't see the sense of letting all the Peenemunde know-how and expertise fall into the hands of the Russians, the scientists and engineers there received a blanket offer of transport to the United States with their families and a promise of high future civil service status. Toftoy had no actual authority to make this offer but the timing was so tight and the need so obvious that it was the only thing to do. Since the Germans were none too anxious to fall into Russian hands, all who could accepted the offer and wound up in a sort of concentration camp in Fort Bliss, Texas while the Army decided what to do with them. Around a thousand engineers, scientists and their families were shipped over to America. Among them was General Dornberger, commanding officer of Peenemunde, Wernher von Braun who was the technical director and many other outstanding technical people including Scheufelen.

I was given the job of directing the new project which was called Loki, for the Norse god of mischief. My first move was to make Bob Flanagan full-time man in charge, since I still had a number of other projects to manage. Bob and I decided early in the game that we had better get down to Fort Bliss and see what the Taifun boys had to say.

Scheufelen, who spoke excellent English, introduced us to the other engineers of his Taifun team: Dobrick, Neuhofer and Dohm. It became immediately obvious that Scheufelen knew more about the problem than we would be able to learn in the next six months so we resolved to move the four Taifun team engineers up to Teterboro at the earliest opportunity.

The Army man assigned to monitor the project at Teterboro was Lieutenant M.S. (Bill) Hochmuth. He had been at Peenemunde under Colonel Toftoy and had done a masterful job of gathering the V2 parts and getting the German scientists to the U.S.A. in the face of active competition by the British and Russians. Among his first problems was finding apartments for the Germans and their families; none of the local apartment house owners was too keen to have them in their buildings. Bill Hochmuth waved the flag and got them all into a middle-class section in Maywood where, in about two weeks, they were completely assimilated. Dohm played the violin in the local orchestra; Herb Dobrick, whose hobby was watch repair, fixed watches for all his neighbors and many Bendix employees; and Klaus Scheufelen's attractive wife sang in the church choir. Ours were the first four of the so-called "paperclip" personnel to leave Fort Bliss and go into private industry. The rest were gradually absorbed; Dornberger wound up at Wright Field working for the Air Force and von Braun went to Redstone Arsenal and later became head of NASA.

Because of the highly informal nature of the arrangements for bringing the "paper clip" group into the country, the State Department had to ship them all across the Mexican and Canadian borders so that they could re-enter legally and pass through immigration. In the case of our four, the Army had no known mechanism for paying them after they left Fort Bliss, so we had to convince Bendix, against all its principles, to advance their salary with the vague hope of being reimbursed some day. Largely as a result of Bill Hochmuth's diplomacy, we got through all these hurdles and were able to form an efficient organization.

This was the largest task I had tackled so far. My first job was to estimate the budget and organize the personnel and facilities. Although I leased a disused hanger on nearby Teterboro airport as headquarters for Loki, I maintained my office in the Bendix engineering building for other projects. Loki engineers, draftsmen, procurement, secretarial, mathematicians, instrumentation and propellant technicians were all located at the airport, including the ex-Peenemunde personnel. Later we took more space on the ground floor of the hangar for assembly of experimental and prototype missiles. For static testing we were assigned a site at Picatinny Arsenal near Reaction Motors, and I hired an architect to design the test cell under the direction of Herb Dobrick of the Peenemunde group.

The Loki design team included fourteen draftsmen under Maurice Vauclair, mathematicians for ballistics and aerodynamics, seven instrument engineers and technicians, twelve static test personnel under Frank Gaffney, and various other miscellaneous personnel for a total of ninety-two people. In addition we used the facilities of the Eclipse machine shop and the part-time help of such plant experts as Ken Halliday for fuels and others for metallurgy. We also made much use of the experts in ballistics and explosives at the Aberdeen Proving Ground in Maryland.

The missile itself was unguided and was designed to be fired in clusters from a 64-tube launcher, all 64 tubes being fired nearly simultaneously to create a shot-gun effect. Because of the very short burning time (less than one second) and the high final velocity (Mach No. 4), the missile was calculated to stand a statistically valid chance of intercepting an enemy aircraft regardless of evasive maneuvers, assuming that the tracking and fire control systems worked correctly. The missile consisted of a booster section of about 5-1/2 inches diameter and five feet long with tail fins and a conical receptacle in the front end, into which the conical rear end of the warhead fitted. The warhead was pencil shaped, about 1-1/2 inches in diameter and three feet long with its own tail fins and contained a couple of pounds of explosive. As soon as the fuel in the booster was exhausted it would separate from the warhead which would continue toward the target while the booster fell back to the ground. Propellants were again red fuming nitric acid for the oxidizer and xylidene for the fuel, with a small squib to pressurize the tanks and initiate combustion. A self-destruct feature was built into the warhead in case the target was not hit.

The Loki configuration was an improvement over the original which Klaus Scheufelen had developed at Peenemunde but it was to the advantage of the Allies that the Taifun never got into full production. Klaus had many interesting stories about life at Peenemunde which had its share of internal politics like government and industrial organizations everywhere. They had several other potentially effective weapons such as the Wasserfall, an anti-aircraft beam rider. Most incredible was the Natter which was piloted and designed to be launched on a vertical rail fastened to the side of a straight pine tree. Power was supplied by a Walther hydrogen peroxide and fuel engine. If the pilot was lucky enough to survive the take-off (many of the first prototypes exploded), he would shoot straight up into an Allied formation, firing rockets at the bombers and then presumably parachuting out at a lower altitude. These rocket weapons developed at Peenemunde were called "Vergeltung," translating to "reprisal" in English. Because of an almost complete lack of hard liquor in Germany by then, every test cell had its own still for making potable alcohol out of the denatured variety used as fuel in the V2, the resulting mixture being called "Vergeltung Schnapps."

A large number of explosions in our static tests were expected until we could generate satisfactory ignition and injection characteristics. No one was too surprised or disappointed when a lot of boosters blew up. While the rocket development proceeded we also started work on the 64-tube launcher. A big project in itself, I decided that the initial concept phase was the sort of thing an industrial design firm could do best so I got bids from several such outfits including WDTA. Actually the WDTA bid was the best in my opinion but Roy Sylvander and I agreed that an appearance of nepotism was to be avoided so we awarded the contract to the Henry Dreyfuss office which did its usual good professional job.

Typical of the many problems that came up was the question of what would happen to the boosters after separation. Presumably, if Loki became operational, with a number of 64-tube launchers in action, the sky would be full of boosters falling like confetti in a Fifth Avenue parade. There was wide disagreement between the aerodynamicists at Teterboro, Aberdeen and the Pentagon on how the falling empty boosters would act. One school was convinced that they would tumble on the way to the ground, others were sure they would stabilize and come down vertically. The answer was quite important since the weapon's use would be seriously compromised if it was going to create a lot of civilian casualties in populated areas. The best way to find out for sure would be to take some boosters up in the air and drop them so I organized a drop test series over the Hudson River at the end of the unused Piermont Pier which sticks out about a mile into the Tappan Zee near where my brother Lewis and I had our duck blind. This required all sorts of approvals from various local, state and federal agencies including the Coast Guard and the Corps of Engineers, but Bill Hochmuth's usual diplomacy finally got us our clearance. We made up a number of dummy boosters, correct as to configuration and weight, plugged so that they would float and the Bendix Twin Beechcraft made a series of runs over the end of the pier dropping the boosters out the door at various altitudes and attitudes. We borrowed Joe Nero's boat, the one we used for duck shooting, and Bob Flanagan and I acted as the recovery team. As it turned out, all the experts were wrong. Every one of the boosters came down in a perfectly

horizontal position, spinning lazily, at a velocity so low that they could have been caught bare-handed.

After many explosions at our Picatinny test stand, we finally got the booster combustion chamber and injectors to the point where they were igniting and burning reliably, the warhead design was finalized, and we were ready for our first actual firing. I decided to use one of the Army's 155 mm cannons as the launcher since its six-inch bore would accommodate the aluminum launcher tube nicely; it was rugged enough to withstand any possible explosion, it was mobile and it could be aimed easily and accurately. The first firing was to take place at the Aberdeen Proving Grounds on Chesapeake Bay. The Aberdeen people were understandably very cautious with new weapons so the first firing was set up to shoot horizontally at an enormously high stone wall, at least ten feet thick at a range of only about 300 yards so there could be no possibility of missing the wall. So that no one would be near the missile after both propellants were in the tanks, we devised an automatic remote control loader which was attached to the 155 mm gun. Nitric acid was already in the missile but the loader filled the xylidene tank and then tipped the missile into the horizontal position and finally rammed it forward into the launching tube. There was an elaborate countdown procedure with each move carefully spelled out so that every possible eventuality was taken into account. This loading procedure took about an hour. When we felt that everything was as ready as it could ever be, we set a date for the first firing, trying to maintain secrecy so that if there was any trouble we could fix it without getting the Army too excited.

A naive hope on our part. When the big day arrived the test site was swarming with generals and the Assistant Secretary of the Army was on hand. In order to play doubly safe the Aberdeen authorities had all fishing boats cleared off that section of Chesapeake Bay for the day. The weather was rather ominous with lowering clouds to suit the occasion. All the Bendix and Army personnel were herded into remote bunkers with narrow viewing slits. The countdown and the filling procedure apparently went smoothly, the automatic loader tilted the charged missile into position and rammed it into the launcher tube. When the big moment arrived, there was a mild pop like a child's firecracker, the missile slid out of the end of the launcher, flopped along

the ground for a few feet and lay there, emitting a large red cloud of nitric acid fumes. The minute I saw this I knew what the trouble was; there was no xylidene fuel in the missile. Our 100% foolproof procedure had gone wrong somewhere.

After the disappointed generals had left for Washington, we went over the missile and loader carefully. It didn't take long to discover the problem. One part of the hour-long loading procedure consisted of taking the cap off of the end of the rubber xylidene tube; later on the tube was connected to the missile. Both operations had been carried out according to the book, but in the ten- or twenty-minute interval a mud wasp, searching for a good place to lay an egg, had spotted the uncapped xylidene tube, laid an egg in it and tamped it down securely with a plug of mud, according to normal mud wasp practice. So when the tube was connected, it was completely blocked and no xylidene could enter the missile.

We sectioned the mud wasp nest and tube carefully, mounted it in Lucite, and sent it down to the Pentagon. A couple of weeks later we made our second attempt. This time there was no doubt about whether the Loki worked; the flash of the rocket and the explosion as it hit the wall occurred seemingly simultaneously. One burning piece of the booster flew up over the wall and landed on the far side where it set the grass on fire. The general in the bunker with me exclaimed, "Jesus Christ, it went through the wall!"

Around this time, Scheufelen received permission to return to Germany to help re-activate the paper factory that he had operated before the war, and he took Herb Dobrick with him to be chief engineer of the plant. Our stalwart liaison man, Lieutenant Bill Hochmuth, was transferred to other duties and replaced by a new Army officer. The new man and I hit it off badly from the start; I refused to agree to those of his demands which I felt were unreasonable. There was only one possible outcome of this clash of personalities. I was removed as project director. This was not a big disappointment since Loki had proved itself and I had already launched on a new project which appeared to have very interesting possibilities.

Bendix was still searching for an answer to a practical, on-board, jet engine starter. The problem

Loki's first successful test firing at Aberdeen.

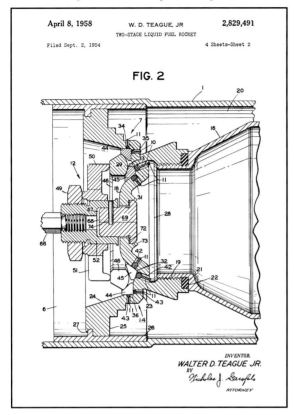

One of Dorwin's three post-Loki patents.

had been assigned to the Research Engineering Department, under my direction. Our first approach was to use concentrated hydrogen peroxide as a mono-propellant, acting on a small turbine wheel geared to the engine. A far different animal from the classic hair bleach, hydrogen peroxide or H_2O_2 in 90% concentrated form is a highly dangerous fluid because once decomposition begins the progression becomes rapid. It decomposes into steam and oxygen at a temperature which is fairly high, but nothing that a steel turbine wheel can't handle, and the result is a lot of energy per pound of fluid. If even more thrust is desired a second fuel can be added to combine with the free oxygen molecule, which is what the Germans did in the Walther engine for the Natter rocket/plane. This bi-propellant version is even more dangerous and the jet is much hotter so we planned to use the straight steam system.

As long as 90% H_2O_2 is not heated up too much, it is safe to store, providing there is zero contamination in the storage tank. But the possibility of human error is always there and one can never be 100% sure of absolute cleanliness. Only a tiny speck of dirt can initiate the inexorable build-up of temperature and pressure and final explosion. At one point we had about a dozen 55-gallon drums of the stuff in a special shed I designed in a remote area of the grounds. The drums were stored over an artificial pond on tilt platforms and each one was constantly monitored for temperature so that if there was any sign of temperature build-up, the offending drum could automatically be dumped into the water and the contents diluted. I remember one day a driver came into my office in the main building with a 55-gallon drum of H_2O_2 on a dolly, dumped it on the floor and said, "Where do you want this, Bud?"

The H_2O_2 starter was not too difficult a development as long as one kept the requirement for 100% cleanliness in mind. Ken Halliday, our fuel specialist, got a bit careless one day and spilled a few drops into his shoe. He could have won a break-dancing contest. An experimental version of the starter worked all right but it finally became apparent that none of the military services was happy with H_2O_2 because of the safety problems. For a while the Navy had used it to propel torpedoes but stopped doing so because of the hazard to submarines which had plenty of their own problems to worry about.

Our next approach was to use a solid propellant like the material used in the JATO units to shorten aircraft take-offs. Bendix had been supplying so-called cartridge starters for years for starting reciprocating engines. However, the jet engine was a much different problem. We estimated it would take twenty or thirty seconds to bring the jet engine up to its self-sustaining speed of around 2500 rpm and there was no known high temperature turbine material that would take the high temperature of the solid propellant. The British were using solid propellant to start their jet engines so Bob Flanagan and I flew to England to visit the Rotax people to see how they did it. Director of Engineering Roy Sylvander gave us strict instructions to get to London, get the information and get back. He didn't want any foolishness about coming back by sea but, as an afterthought, he said if we did sail back we would have to write our reports on the ship. Roy was a great person to work for. Accordingly, Bob and I booked passage on a DC4, which stopped at Gander in Newfoundland to refuel and, for the return trip, we booked first class cabins on the *Queen Mary*, which by 1948 had been restored to her pre-war luxury. This was the first European trip for both of us. We checked into the Savoy Hotel and were entertained royally by the Rotax people and by old friend Freddie Fisher of the Braman-Johnson Flying Service who was now married to Ann Babington and living in Chelsea.

The turbine heat problem was quickly solved. The British merely stepped up the power and burning rate until it brought the engine up to 2500 rpm in about 3-1/2 seconds so that the turbine wheel never had a chance to burn up. We cursed our own stupidity as this simple solution had never occurred to us since we were always thinking in terms of lower torques and longer times. One of the most impressive parts of the tour Rotax gave us was a look at the remains of a test cell in which one of the Walther H_2O_2 plus fuel engines had exploded. The previously rectangular cell was now elliptical in cross section and the bullet-proof glass viewing port had instantly killed all four observers.

This was a useful lesson in how not to build a rocket test stand. It was also a lesson in what can happen in a liquid propellant rocket if something goes wrong. The Walther engine being tested had quite large, strongly-built fuel tanks for the two propellants and both tanks were full, just as they would be at take-off. No one had imagined that both propellants could instantly co-mingle and yet this is just what happened. The exact sequence of events is uncertain but the final result was that the two propellants mixed instantly and efficiently to form an explosive mass which was equivalent to hundreds of pounds of TNT in energy release and was so powerful that it pushed the reinforced concrete floor of the test cell down into the ground a couple of feet. Even a properly designed test cell wouldn't have helped much.

A similar thing happened in the Challenger disaster which took place in a completely unconfined situation, and yet the force of the explosion of the two propellants was so great that it tore the attached, sturdily-built spacecraft to bits. Familiarity tends to erode respect for the potential hazard of hundreds or thousands of pounds of two elements of an explosive mixture (in exactly the correct proportions for maximum effect) in close proximity to one another. Large rockets are no place for school teachers; it will be a long time before their reliability approaches that of other means of travel, regardless of how careful and conscientious the cognizant development group may be.

We had time for a two-day visit to Paris where I attended my first Catholic church service at Notre Dame, plus the usual round of nightclubs and gourmet restaurants. Back in England Bob and I spent a couple of days in Stratford on Avon where we stayed in a beautiful chalet which had been converted to a deluxe hotel. We also hired a car to watch the Bugatti Owners Club hill climb at Shelsley Walsh. The trip back on the QM lived up to our expectations. The maitre d'hotel in the first-class dining room challenged us to think of a dish he couldn't supply so one night I ordered wild partridge, which was delicious and still had a couple of #6 shot in it. Another example of Cunard style was the live oyster bed down in the lower bilge

European trip, 1950. Bob Flanagan and Dorwin's friend Freddie Fisher in London. The chalet/hotel in Stratford on Avon. Watching the Bugattis at Shelsley Walsh.

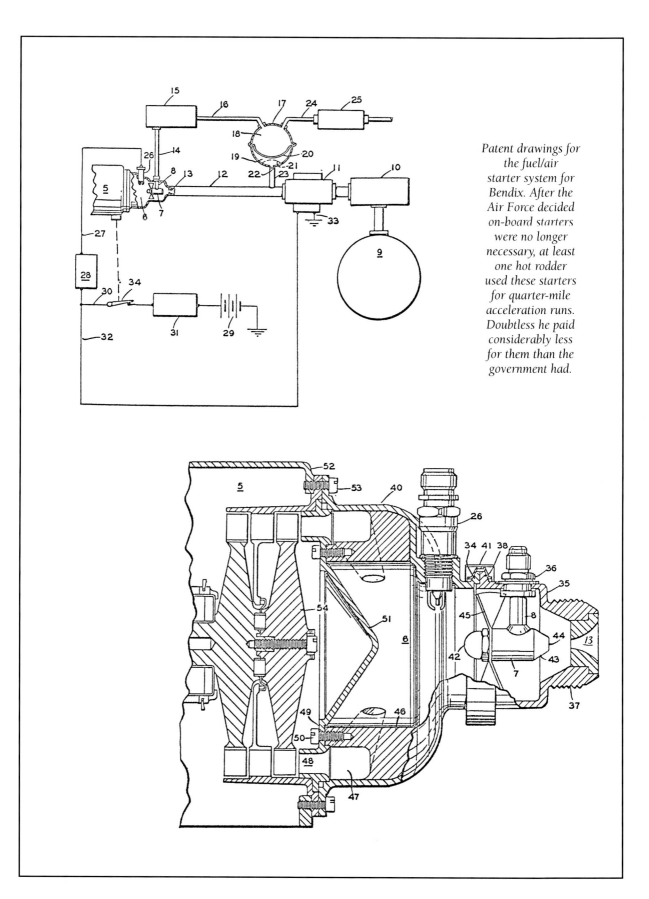

Patent drawings for the fuel/air starter system for Bendix. After the Air Force decided on-board starters were no longer necessary, at least one hot rodder used these starters for quarter-mile acceleration runs. Doubtless he paid considerably less for them than the government had.

section so that passengers could have fresh oysters with their meals. I think Roy Sylvander realized that the passenger-liner era was nearly over. I'll always be grateful to him for letting us have the experience.

Back in Teterboro we started work on a solid propellant starter system but this solution didn't thrill us. It was not as bad as H_2O_2 but still rather unsafe. A huge charge was required and it was a logistic problem to have this propellant always on hand. Solid propellant doesn't store well; if it gets too cold, it may crack due to vibration which can accelerate the burning rate to a catastrophic degree. I don't know where I got the idea but one day it occurred to me there was a propellant combination every jet aircraft already had on board: the standard JP4 engine fuel plus air, which is 21% oxygen, both of which are available in unlimited quantities. At first I wasn't too serious about it but I ran some rough figures and the more I thought about it the better it looked. Together with Al Volk of our department who was better at turbine calculations, we started putting together a preliminary design. As far as temperature was concerned, the nitrogen and carbon dioxide in the air diluted it somewhat so that if we stuck to a three- or four-second starting cycle the turbine wheel would take it all right. The figures showed that a teacup full of JP4 would work for a start and a sixteen-inch spherical tank of air, compressed to 3000 psi, would be adequate for the three starts in the jet engine we were concerned with. A 3000 psi compressor driven by the engine could recharge the air tank in flight so there were no longer any logistics problems with dangerous propellants.

One possible hazard we recognized early was the 3000 psi air tank itself, which on a military aircraft might create shrapnel if hit by a 50-caliber machine gun bullet. Fiberglass was just beginning to be used for boats in 1948 and it seemed to me that a fiberglass sphere would not only be strong and lighter than metal but probably would not shatter if hit. We contacted one supplier, I think it was Union Carbide, who constructed what I believe was the first geodetically-wound fiberglass pressure vessel ever made. The first attempts were wound on an ordinary steel globe of the world that happened to be close to the right size but it wasn't successful as the expansion rates of the steel globe and the fiberglass were different so they finally wound the glass on a wax sphere which was melted out after the winding was complete. I don't remember if the plastic resin was polyester or epoxy but it was air tight. Several prototype tanks were made and the Air Force ran tests with tumbling 50-caliber machine gun bullets. The bullets merely made a large hole; the glass fibers fluffed out but there was no tendency to shatter.

When this development began to look promising I had a special starter test shed built alongside the altitude chamber. We finally settled on a turbine wheel of about seven inches in diameter, running at 65,000 rpm at the end of the cycle. This was geared down to about twenty-five to one to a heavy-duty Bendix drive which engaged with the engine. Ignition was by a standard spark plug and a regulator controlled the flow of JP4 to insure a stoichiometric fuel-air ratio. Machined turbine wheels were quite expensive so we decided to experiment with cast wheels. The first one we tested blew up at a relatively low speed. A member of the test crew was leaning over the unit at the time which was against orders. One piece just ticked the end of his nose, other pieces stitched a row of holes across the building's roof and wall, and another piece was found in Hasbrouck Heights, a quarter of a mile away. After this everyone paid strict attention to safety rules.

Al Volk and I got a joint patent on this starter which eventually became big time for Bendix. A special plant was built in Utica, New York to produce the starters, and many millions of dollars worth were sold to the Armed Services.

During my time at Bendix my brother Lewis was racing outboards in the summer. I would occasionally enter some local race when I had the time. In 1948 the American Power Boat Association decided to change the rules for the Hudson River marathon—an annual race starting in Albany and finishing at Dyckman Street or 79th Street in New York City, a distance of 140 miles—to admit only displacement hulls and stock outboard motors. The race had always been run with hydroplane racing hulls and racing motors; the new rules were meant to encourage more entrants—anyone with a fishing boat and motor could enter. Our friend Vic Scott, who was always up with the leaders and had won the overall prize several times, got us interested in the new set-up so we decided to enter, Lewis in Class C, me in Class A, the smallest class. Another friend, Bob Ackerley, was a dealer for Feathercraft Aluminum boats and agreed, with Feathercraft's blessing, to supply us with the boats. Lewis already had his motor. George Shortmeier, son of the distributor for Champion outboard motors in New York, agreed to give me a motor for Class A.

Despite a big turn-out for the race Lewis and I had a number of advantages. For one thing we were able to practice in the roughest part of the river, whereas Midwest entrants had no experience with the Hudson, which gets very choppy when the wind and current are opposed. Also we had Vic Scott's long experience which included legal but little-known tricks like saving some distance by going behind Bannerman's Island or having a typed route

Lewis and Dorwin with Lewis' boat at the Alpine landing and practicing for the race.

description taped on the deck in front of us. Certain modifications were within the rules such as a special propeller or adding a little castor oil to the fuel mixture, all of which we took full advantage. Because of the distance, everyone carried a large fuel tank in the boat from which to periodically fill the motor tank by means of an aircraft wobble pump and fuel resistant hose.

In the 1948 race Lewis won Class C and Vic Scott, as usual, took overall honors. I didn't do so well because I had installed a tank large enough to avoid refueling. With the extra weight and the small motor, the boat didn't really get going until the race was half over. Lewis's prize was the huge ugly Hearst trophy that he had to lug up to his fifth floor walk-up on Spring Street and even had to pay insurance on it.

For the 1949 race I had learned my lesson on weight and went all out to make my outfit as light as possible. I got a very light aircraft fuel tank which held just enough to get me to Poughkeepsie, the halfway point. I even made my own automatic shut-off throttle which was much lighter than stock throttles. I adapted an aluminum aircraft steering wheel and made up four special lightweight steering pulleys using Micarta aircraft sheaves. The Bendix machine shops were always willing to work on personal equipment. These were known as "government jobs" and were tolerated as long as they were kept within reason. A beautiful tapered high strength alloy steering bar was built for my motor that must have saved one or two pounds. I also designed an extremely light boat that Feathercraft made up for me. The first test in rough water showed that it wouldn't last the route, however, so just before the race I borrowed the same boat I had used the year before and transferred all the gear to it. Frank Gaffney from Research volunteered to refuel me at Poughkeepsie and we practiced the routine until we could do it in a few seconds.

Lew and I always got good starts because we had done a lot of racing. This was important in 1949 as there were three hundred boats starting in two heats and the river is only a few hundred yards wide at Albany. As soon as I was clear, it was obvious that the boat was going well and I made good time until about twenty-five miles from Poughkeepsie when I noticed fuel sloshing around my feet. In my haste to convert to the new boat, I hadn't secured the fuel tank adequately and it moved forward until a screw punctured the hose to the wobble pump; all the rest of the fuel was in the bilge.

Because I had just pumped the motor tank full, I knew I had a few miles before I would run out. However, I also knew that if I stopped at a gas dock, by the time I bought gas and oil, mixed it up and got it in the tank, I would have no chance of

The two brothers with Dorwin's Feathercraft outboard. Dorwin's J3 being lowered for battle.

winning. My only chance would be to find a fisherman with a spare fuel can. Spotting a pair of them anchored over near the west shore, I yelled and asked if they had any spare fuel. One of them held up two fingers indicating two gallons. I slammed the boat alongside and cut off the punctured end of the fuel hose which was long enough so that I could make it up again, then I forced the tank back into the proper position and wedged my fire extinguisher into the space to prevent it from moving forward again. In the meantime the fishermen (may they always be lucky!) poured their two gallons into my tank. I put the cap on, gave them a few dollars, and roared off. The whole operation couldn't have taken more than two minutes.

I reached Poughkeepsie all right and Frank refilled my tank in record time. It was quite rough on the southern part of the river where my larger boat was an advantage in the choppy water. As I crossed the finish line I spotted Lewis dancing on the float and holding up one finger. I had won Class A.

The Class A prize was $500 in cash. I was afraid it might affect my amateur standing in sailing. The committee suggested a compromise whereby the prize would be credited to any store I chose. This was rather a fine distinction but I agreed and selected Sears, Roebuck. The amount of merchandise one could get at Sears for $500 in 1949 was quite amazing. In addition to a band saw, a drill press and a lawn mower for myself, I was able to give nice presents to Harriette and son Harry. Champion Motors and Feathercraft were both delighted with my win which was worth a lot to them in publicity. They gave me the boat and the motor. Lewis had not placed in the first row this time, but our friend Vic Scott had again won Class F and the overall trophy.

In the fall of 1949 my brother contracted polio. In an iron lung for several weeks, Lewis was finally moved up to the Haverstraw, New York rehabilitation center. He had been doing a lot of painting, mostly in an abstract style, and was beginning to be recognized. Left-handed from childhood, Lewis lost the use of his left arm completely, but his right arm was able to perform some functions. He recovered partial use of his legs and could walk but had to have a special chair to be able to get to his feet. His marriage had begun to break up before his polio attack and ended during it. In spite of his misfortunes, he never lost courage or his keen sense of humor. After his release from Haverstraw Lewis taught himself to paint with his right hand. He married Virginia Vanderbilt from Seattle, and they moved to a house on Dick Jewett's estate in Upper Nyack on the Hudson where their first children were born. A few years later they moved again, to Norwich, Vermont, across from Hanover, New Hampshire. The polio attack was especially hard on Lewis because he had always been very active in skiing and other sports. He transferred his love for sports to his children, becoming active in the U.S. Ski Association and helping organize an Austrian instructor for the local Norwich kids. Of his four children, Alison, Cece and John became good enough to be picked for the U.S. Ski Team. Cece won the Roche Cup in Aspen twice and John was a member of the University of Vermont ski team when they won the national intercollegiate championship in 1980.

I entered another Albany-New York race in 1950 but one of the reed valves in my motor broke and I had to drop out near Catskill. It wasn't as much fun without Lewis so I got rid of my outboard stuff and bought a Thistle class day sailer, the first on the Hudson River. Harriette and I raced with the local lightning fleet and one summer we put the sailboat on a trailer and took it up to the Brick House to sail on Winnepesaukee.

Page opposite: Dorwin winning Class A in the Albany-New York, 1950. A happy Harriette in the crowd. The Bendix party for the Class A winner, Klaus Scheufelen and Herb Dobrick on the far right. This page: Albany-New York, 1950, the ignominious end. A sketch Dorwin made during his brief outboard racing days.

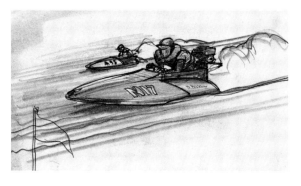

Chapter Ten

1951 – 1952

Every time Harry fell, Lewis purposely took a tumble himself ...

Throughout this book I have put a lot of emphasis on sports—sailing, skiing, racing cars and outboards, etc. With all the time I spent on extracurricular activities how was there enough left for serious design work? Actually, these "frivolous" activities made it possible for me to get as much work done as I did. The list of projects I completed at Bendix, WDTA and later in my own company is long. Without the relaxation that these outside pursuits provided, I would never have been able to sustain this level of creative effort. Actually, the line of demarcation between vocation and avocation has often been blurred; my sailing led to profitable boat design, skiing to lucrative work on skis, boots and poles, car events to automobile design and articles. In addition, participation in these sports gives an awareness and feel for ergonometric factors that instinctively leads to the proper design direction. Ford's ascendance in design and sales can be partially traced to chairman Don Petersen's directive that his top engineers and designers go through Bob Bondurant's driver school course.

Of all sports, sailing is the one most suited to industrial designers. It has all the advantages of other sports—good exercise in the open, perpetual search for perfection, occasional elements of risk, and generally beautiful surroundings. In addition sailing involves equipment—the boat—which offers unlimited scope for devising improvements, making

Dorwin's interest in sports is virtually all inclusive, as these sketches indicate. He entitled his baseball sketch "Foul Ball." The boxer and the tennis player sketches were unnamed.

gadgets and buying exotic accessories. In fact most boat owners, regardless of their background, automatically become designers in their own minds and dream up all manner of more or less effective modifications. There is a fascination to sailing that grips people of all ages and classes; housewives, bankers, carpenters, insurance salesmen, et al. A group on the West Coast made up principally of aero-space engineers and physicists publish an annual report called *The Ancient Interface* which proposes all sorts of far-out boating concepts. One of their crazy ideas was the now-ubiquitous sail board. Another group, the Amateur Yacht Research Society, based in England (Prince Philip is its sponsor), is largely responsible for the present popularity of multi-hulls and the proliferation of solo and two-man ocean racing. Any stock boat offers vast opportunity for ergonomic and aesthetic improvement; this is the designer's stock in trade so it isn't surprising that many industrial designers take to sailing at first opportunity.

In 1951 our neighbor John Halsted sold his "S" boat and bought Bambi, a 34-foot Hinckley Southwester auxiliary sloop. After Harriette and I sailed a few times on Bambi with John and Nancy, we were thoroughly sold on cruising as a way of life. Fortuitously, about this same time my mother decided to distribute some capital to her children. With part of my share, plus the sale of our Thistle, I bought Sea Frolic, a 34-foot Shuman single-hander, sloop rigged, with a Gray marine engine for auxiliary power. Sea Frolic had been built by Graves in Marblehead; other boatyards admired the solidity and workmanship of her Honduras mahogany planking. Relatively narrow in the beam she still had plenty of room for Harriette and me, with Harry in the forward cabin along with a pair of hamsters and Birabongsi, our handsome Siamese cat, named after the Prince of Siam who was a well-known racing driver at that time. Bira loved the boat and, except for occasionally sharpening his claws on the sail cover, was very well behaved. The hamsters had a wire treadmill wheel in their cage which could be heard spinning all night long.

We belonged to the Nyack Boat Club, a flourishing organization today with hundreds of

Dorwin working on Sea Frolic at the Nyack Boat Club. Sea Frolic en route to Nantucket.

members and many good-sized boats. The dues were $6.00 per year; some controversy followed when they were raised to $11.00 to pay for hurricane damage to the dock and floats. There were only four or five cruising sailboats on the entire Hudson River in 1952 and relatively few in the Sound and the New England waters. The owners got to know each other and their boat and yacht clubs welcomed visiting boats from other clubs. We began to spend weekends and vacations on Sea Frolic, flying the Nyack Boat Club burgee and moving from one club to another; Harriette and Harry stayed on board while I commuted back to work by car, ferry, train or air. We visited Larchmont, Seawanhaka, Shelter Island, Montauk, Watch Hill, Newport, Block Island, Padanaram, Cuttyhunk, Edgartown, Nantucket, et al. The first time we came into Nantucket Harbor in 1952, the Nantucket Yacht Club tender came alongside with a guest card and an invitation to a dance that evening, and then showed us to a free guest mooring. Each new harbor was an adventure, everyone was helpful, and we made many new friends. Sea Frolic was a handsome boat; we stored her for the winter in Arthur Knudsen's yard in Huntington, and kept her in good condition. Three decades later I was on the ferry from New Bedford to Vineyard Haven, admiring a very smart, black-hulled sloop with new sails when I suddenly realized from the sail number that it was our old Sea Frolic. Nice to know that she was well cared for and appreciated.

For his seventh birthday I built Harry an eight-foot plywood sailing pram from a kit put out by Chris Craft. (We needed a new dinghy for Sea Frolic anyway.) When I went back to New York to go to work, Harry asked permission from his mother to take his boat out alone. Because we were moored at Watch Hill, a very snug and secure harbor, Harriette agreed. But Harry sailed to Stonington and back on his maiden voyage, a total distance of about seven miles; his mother was quite relieved when he returned several hours later. After this, every time we reached a new harbor, Harry would rig the sail on his dinghy and sail around the other boats where he often was asked aboard. Later he would introduce us to the new friends he'd made. We got to know many interesting people that way.

During the winter of 1951-52 my brother-in-law, Arthur Meurer, lived on Manhasset Bay. This was one of his prosperous periods. His rented house was next to the Manhasset Bay Yacht Club but was larger and more imposing so that many people mistook it for the club. Arthur had chartered an elaborate floating duck blind which was moored on the other side of the bay. Before dawn one very cold morning in late January my young nephew John Meurer and I started out for the blind in Arthur's outboard. Because it was pitch dark and there was a thin layer of ice on the salt water, we proceeded slowly and cautiously. About halfway across the bay I suddenly realized there was water in the boat. John was at the wheel forward and I was sitting aft; I told him we were in trouble, to open the throttle wide and head for the blind as fast as possible. Manhasset Bay at 5 o'clock in the morning in 15° weather is no place to be with a sinking boat. I bailed like mad and we made it all right. Hauling the boat up on the spacious deck of the blind, we found that the thin film of ice had cut a neat four-inch gash in the 5/16 mahogany planking of the hull. Opening up the motor raised the hole above the water most of the time and slowed the amount of water coming in.

After we dried out the guns and ammunition as best we could, we turned the boat upside down and made a temporary repair with tar paper, lathes and nails that we cannibalized from the blind. As I remember we had fair luck with the ducks and the repair held up all right for the return trip.

During the winter months Harriette and I went skiing every weekend. By the age of five, Harry was good enough to come along. Before my brother Lewis contracted polio, he taught Harry turns and technique; every time my son fell, Lewis purposely took a tumble himself to show that it didn't hurt. Lewis and Harry became great friends; Harry often visited with him in Norwich and later skied with Lewis' children at the Dartmouth College skiway.

West Coast visits to Lockheed and Douglas on Bendix business added to my ski sites. On my first trip to Aspen, in 1950, I had reservations at the Hotel Jerome, but in Denver found out that the roads to Aspen were closed due to heavy snowfall and avalanches. I booked a flight to Grand Junction to approach from the other side but the roads were no better. So I chartered a small, single-engined plane to fly in. We got to Aspen all right, but the airport was covered with several feet of snow. After circling the valley a few times, we were about to give up and return to Grand Junction when a car stopped on the main road, which was plowed, and the driver jumped up on the bumper and signaled us to land on the highway, which we did. The car had been sent out from the hotel so I was all set.

Aspen had just put in its first single chair lift up Ajax, at the time the longest lift in the country. On my first ride up I stopped at the glass-enclosed warming hut at the top, where a couple of ski patrol types were shoveling in a nearby snow drift. As I watched, they uncovered two one-armed bandits. Getting word of these illegal slot machines on the previous day, the sheriff and his deputy had donned their ski clothes and bought lift tickets to make an unannounced raid. The attendant at the halfway station had recognized them, however, and waited until they were over a 100-foot drop, stopped the lift and phoned the top of the mountain for the machines to be buried in a nearby drift. As soon as this was done, he restarted the lift. On my first trip down the mountain I was all alone and the snow was unmarked with a few inches of powder over a deep base. It was all so perfect that I was yelling for joy as I zig-zagged around the pine trees on the way down.

On another trip I had reservations at Alta in Utah. Again the road from Salt Lake City was closed halfway there because of avalanche danger. All the other cars were turned back but the road patrol said our Alta Lodge station wagon could go through if we wanted to take the chance. We voted unanimously to go. We got to the lodge but the

Teagues learn to ski early. Son Harry on his first skis. Nephew John, with whom Harry skied, a Can-Am Slalom champion at age nineteen.

snow avalanched behind us and no one could get in or out for four days. This turned into one of the biggest and wildest parties I've ever attended but we were able to do a lot of powder skiing (which I never really mastered very well). I had to bring back the Salt Lake newspaper accounts of the avalanche to explain my tardy return to Bendix.

While in his office one day, Sylvander and the general manager of Teterboro were discussing the possible acquisition of Metalwerke Plansee, a new high-tech powder metallurgy plant in Reutte, Austria. Before making a decision they wanted to get more first-hand knowledge and wondered who they should send over. Naturally I said I would be glad to go. I also convinced them that, if Harriette came along, we would be able to get better inside information into the political leanings of the Metalwerke executives. Reutte is located near Garmisch Partenkirken in the heart of the best Austrian ski country. The trip was arranged to coincide with the height of the ski season.

By this time WDTA was working for Boeing in interiors and had done the Stratocruisers which were large four-engined propeller planes with a bar and lounge on a low deck, reached by a spiral staircase—sort of a 747 upside down. We booked our flight to London on a Stratocruiser. In those pre-terrorist days interested passengers were invited to come forward into the cockpit and sit in the captain's or the co-pilot's seat while they visited in the main cabin. After a pleasant twelve-hour flight, with lots of champagne, we landed in London, checked into the Savoy and spent a couple of days touring around the city and the countryside with the Fishers. Flying to Innsbruck, we took a train the next day to Zurs in the Voralberg where we had reservations at the Alpen Rose. It was crowded and noisy so we quickly moved over to the Lorunzer Hof which was owned by Count Skardarazy, who had come over to the USA with Ludwig Bemelman when he first worked at the Plaza. Apparently I had a penchant for avalanches because it snowed continuously, day and night, for the next week, the worst season ever experienced in the Tirol. During the night from time to time we could hear the avalanches rumbling down the mountainsides. Every morning guests who had driven up would have to probe with ski poles to find their cars. The first day at the Lorunzer we took a bus to Lech. A small avalanche covered the road before we started back and both bus crew and passengers shoveled furiously to clear a path wide enough to blast through. No one could leave Zurs for several days. All buses in the Tirol are equipped with shovels and oxygen bottles in case they get trapped. In Zurs one skies on the Zurseree in the morning and on the Hexenbogen on the other side of the valley in the afternoon, so as to always be in the sun. As a result of the weather we were late for our appointment in Reutte. But there was a congenial group of British and Austrians at the Lorunzer Hof, the food and accommodations were superb, the skiing was great and everyone enjoyed the enforced stay. Finally the roads were cleared enough so that we could catch the train at Langan for Reutte.

The president and owner of the Metalwerke Plansee was named Schwartzkopf. I had met him in Teterboro but now I saw him in his native Tirolean costume, which is more practical for Austrian winters. The plant was the most beautiful factory I had ever seen. Inside it was modern, efficient and spotless, well lighted and comfortable. Outside it had been laid out to fit in with its surroundings on the Tyrolean Alps. A ten-meter-wide trout stream swept down the mountain next to the slope which was lighted at night so that employees could practice slalom after work. Most of the executives, including Dr. Schwartzkopf, lived in chalets scattered in the woods on the far side of the trout stream, reached via bridge. The whole complex was a lesson in how to design an industrial plant that is compatible with its environment. It must be a pleasure to work in such a place. While we were being entertained, Harriette mingled with the wives and got lots of gossip on the politics and habits of the various executives. I was able to prepare a good report on my impressions of the operation, which were generally favorable.

I would spend quite a bit of time in Austria and Germany, much less in French-speaking countries. For quite a few years in school I studied French and, with a little practice, could be quite fluent, but never studied German. I have a good ear for pronunciation, however, and I picked up a smattering of German from my association with Scheufelen's group. Harriette did study German for a short time and still had some vocabulary so she would supply the German word and I would give it the proper guttural intonation. This, coupled with the fact that most upper class Germans speak good English, got us by with few problems.

On my return I decided that ten years at Bendix was enough. I wanted to get back into industrial design. By this time WDTA had been reorganized as a partnership with my father as senior partner and Bob Harper as the number two man. It was agreed that I would return as an associate and be in charge of Navy inspection work which had grown into a substantial business from the war years. I would also be able to get back into some product design work, most of which had been handled by Bob Ensign after I left. Bendix gave me a rousing send-off party at Petrullo's Everglades and a parting gift of a beautiful Colt Woodsman target pistol, something

I had always wanted.

I've always been glad that I had this experience with Bendix. Ten years is a big part of one's life but I was able to prove to myself that I could get along nicely on my own merits with no help of any kind from my father. I started as a junior engineer at half my WDTA income but was able to live within it and to graduate to head of the Research Engineering Department which I built up from a small group to over 100 people, with substantial annual profits from work with the government and private aircraft companies. I was instrumental in getting Bendix into rockets, ram jets, turbo jets and associated accessories and I assigned thirty-three U.S. patents and many foreign patents to them, a few of which turned out to be very valuable.

From the family scrapbook. Lewis and Virginia in their MG TD. Dorwin and sons in front of the Teague Ford woody wagon. The oldest and the youngest, T holding Harry. Lewis and Harry with cousin Cammy. Her pet crow; on walks he would fly to a telephone pole, wait for her to catch up, then fly to the next one.

Chapter Eleven

1953 – 1957

The Count knew all the right places to go provided someone else picked up the check …

Coming back to Walter Dorwin Teague Associates after being away for ten years was a strange experience. The office at 444 Madison Avenue now spread out over three floors and during my absence each partner and associate had carved out his own niche. My arrival naturally aroused a lot of misgivings. One of the associates, Gordon Peltz, was in charge of engineering and had evolved a system for inspection for the Navy during the war. He was about to retire and devote his time to the Christian Science Church, so the Navy inspection contracts became mine, plus whatever other jobs came long. Bob Harper, number two man, was a pleasant, easygoing person, very devoted to my father and quite happy to follow his lead. A Cornell architectural school graduate, he had a good conservative design sense, especially in furniture and interior design. Bob Ensign, who had pretty much worked under my direction before I left in 1942, was a full partner now in charge of product design. Frank del Guidice, also a partner, was in charge of the Boeing work and stationed in Seattle. Associate Carl Conrad worked mostly on exhibit design. Another associate, Milton Immermann, was a sort of account executive. Jack Brophy, my father's canoeing partner in the early YMCA days, was comptroller.

The structure of the WDTA organization was very loose but it worked all right. By 1953 it had grown to about 200 people and was probably the largest industrial design organization in the country. A year after my return, all of the associates, including myself, were made full partners with equal 8% shares of the profit, except for Bob Harper who had 18% and my father who had 42%. Later on I was raised to 11% so that I became number three in the hierarchy. Jack Brophy was also made a partner but at a slightly lower share of the profits at 6%.

Since most of the Navy work was done at the various sites where the inspections took place, WDTA had a small staff at several locations in California, Indiana and New Jersey. I spent a great deal of time in Washington, D.C. meeting with the cognizant Navy people and visiting the various offices which were located at large aero-space contractors. The best brain on our Navy staff was Sy Gollin, head of the Pomona office, so I promoted him to head the entire Navy operation which left me more time to work on product design. The Navy work was interesting at first but not nearly as fascinating for me as product development and design.

My first design job was a new .22 rifle for the Remington Arms Company in Bridgeport, Connecticut. Guns had always been one of my hobbies. Over the years my collection had grown to around forty rifles, muskets, shotguns and handguns, mostly antiques. I had restored several old guns and had made special stocks for a couple of modern rifles. This is a very exacting art as it involves accurate sculpture in woods like burl walnut that are difficult to work. The metal parts—the barrel, breach and action—must be let into the wood with no visible gap between wood and metal. For me, the Remington job was sort of like the

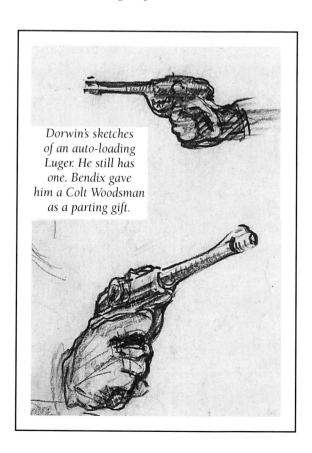

Dorwin's sketches of an auto-loading Luger. He still has one. Bendix gave him a Colt Woodsman as a parting gift.

Marmon Sixteen task in that I had given a lot of thought over the years to what my ideal .22 rifle would be, just as I had my ideal car in mind before I started the design of the Marmon. Remington was pleased with my concept which served as the basis for the Remington "Nylon 22," a best-seller for many years.

Another design and development project was a folding door for a company named Columbia Mills of Syracuse, New York. Its principal business was making coated fabrics for book bindings and folding doors. The Columbia people had decided to make the whole door themselves provided they could find a mechanism that didn't infringe anyone else's patents. Bob Ensign was already working on folding doors for another client, the Hough Shade Company, so I was given the job of developing a new mechanism for Columbia Mills. I applied my usual system of first determining the important objectives, one of the most important being ease of installation. Thus far all of the better folding doors employed a multiple steel pantograph arrangement, to which the fabric was attached at intervals. An overhead track supported the moveable end of the door which otherwise would sag to the floor. This meant the fixed end of the door had to be screwed to the door frame and the overhead track cut to size and also screwed to the overhead part of the frame.

How much easier it would be to install, I thought, if a supporting track wasn't necessary because the moving end would support itself. Then it occurred to me that if there was only one large "X" member, pivoted to the two ends of the door at the top, and sliding in the end members at the bottom of the X, the movable end would be supported rigidly, would no longer have to hang from the overhead track, and would always be held in a vertical position. Well and good, but how could the folded fabric be supported in between the two ends? About this time the light dawned, like the little stylized light bulb cartoonists use to denote an idea being born. Putting the upper half of an intermediate X member between the top halves of the main members provided a point at the top which travels horizontally and can be used as an intermediate fabric support. Repeat this again for three intermediate points, which were probably enough, although seven intermediate fabric supports were available by doubling up again.

At this point, I brought a new employee, Ben Stansbury, into the project. Ben was the son of a head hunter who had contacted me at Bendix about a job for a West Coast company. I didn't take it but we had become friends and I wound up hiring his son. Ben and I now began to think how nice it would be if, using my new mechanism, we could devise a system whereby the door would be installed with no screw or tools whatsoever. The answer we came up with was a spring-loaded extensible overhead guide that did not support the movable end of the door but guided it laterally. At the fixed end of the door, the spring-loaded guide pushed on a ramp so that it exerted a diagonal force against the corner of the frame which jammed the fixed end of the door and the overhead guide in place. We built a

Dorwin's sketch and the Columbia folding door.

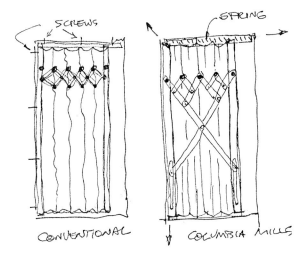

119

prototype of our scheme that worked perfectly and received patent No. 3,065,786. Columbia Mills finally decided to stick with the fabric business and sold the patent rights to our other client, the Hough Shade Company, for many times what our development had cost them.

The door was a purely mechanical development problem. I also tackled primarily aesthetic exercises. The Steinway Piano Company asked us to design a piano to mark its 100th anniversary. I worked with Bob Harper on this project and was responsible for the legs. These followed a "cubic parabola" taper, theoretically the ideal shape for a cantilevered beam of constant strength since the most severe loads on a piano's legs come from the side loads that result when it is pushed across the floor. I also designed the lid to be flush, or actually set back slightly from the sides. Normally all grand piano lids overhang the sides because of the difficulty in matching the curves but I felt Steinway could cope with the flush top which greatly improved the appearance. Steinway was an interesting company to work with. At that time several Steinway family members were in the top executive positions. For over 130 years the plant had been in Astoria, Queens on a street called Steinway. The company's vaunted quality control is no myth; its product is a reminder that one doesn't have to go abroad for excellence of construction and performance. A lot of handwork remains in a Steinway piano but the company doesn't rule out production methods providing they make a part that is as good or better than can be achieved by hand.

Engineering consultation for Servel Corporation was another task. Servel built absorption-type refrigerators based on a principle developed by the Electrolux Corporation in Sweden. Unlike a conventional electric refrigerator, this system had no moving parts but was a rather complicated array of tubing in which ammonia, heated by a gas or kerosene flame, circulated through an evaporator coil in the cooling compartment. The absorption principle provided a satisfactory cooling system, made ice cubes, etc., just as an electric refrigerator does. Its disappearance in the U.S. was due to the high cost of the elaborate tubing, in addition to safety problems (the gas flame had to be inspected and cleaned regularly so that it didn't extinguish or make carbon monoxide). Absorption-type refrigerators are still used extensively in third world countries and remote areas where electricity is not available. Servel had an agreement with Electrolux in Stockholm for the use of its patents and an exchange of information.

Servel was losing business and trying to think of ways to cut expenses. The annual fee being paid Electrolux for yearly reports on new development which were not being made much use of was questioned. Accordingly a junket to Stockholm, Oslo, Paris and London was arranged for meetings with Electrolux and other European manufacturers who might be able to supply the tubing assembly at less cost than it could be built in the U.S. Four Servel executives took the trip, three of whom had their wives along, which seemed like a strange way to begin an austerity campaign. I was asked along to help evaluate the findings and, since I had never been to Sweden, readily agreed. Stockholm is a city made up of islands. Many of the Electrolux workers commuted by water; the plant supplied free winter storage for the employees' boats. My guide to the town was an impoverished count who was a friend of my Alpine neighbor, Dick Huested. He knew all the right places to go, provided someone else picked up the check. We visited a large bathing pavilion on a nearby nude beach. There were separate dressing rooms for men and women but everyone swam together.

By this time my drinking problem had become serious. I was still able to manage sober intervals for interesting assignments but they were becoming shorter and less predictable. A friend who lived in Closter had joined Alcoholics Anonymous and his example persuaded me to give it a try. I knew instinctively that I needed help. Several periods of remaining sober were required before I accepted the truth. I was an alcoholic and I had a choice: stop drinking entirely or die. This was in 1955 and I haven't had a drink since. Furthermore I don't miss it. My example has helped several members of my family.

Stockholm was followed with a short trip to Oslo, Norway to visit the Norweigan Arms Factory whose principle products appeared to be harpoon guns for the whaling industry. Our chauffeur was a famous ski jumper who was also one of the worst automobile drivers I've ever seen, an unlikely combination. Paris was our next stop. After a couple of days there the rest of the party decided to go to Switzerland. The business part of the expedition finished I decided to stay in my comfortable suite at the Plaza Athenée and enjoy Paris. I don't know if our trip was the last straw but Servel Corporation folded thereafter.

In between business trips, in September of 1954, Harriette, Harry (who was ten by now) and I were cruising on Sea Frolic when I had to return to the office and left them on a mooring in Watch Hill. The mooring's owner came back unexpectedly so that they had to drop the anchor, a 30-pound Danforth. When commercial radio announced that Hurricane Carol was approaching the New England coast, I called the Watch Hill Coast Guard immediately. Hurricane warnings were much later and less accurate in those days. The Coast Guard said not to worry, that Carol was headed out to sea, so I went back to work. But our telephone operator's mother was in Watch Hill; she talked to her on the phone and was informed that the Watch Hill Beach Club had just been carried away. I jumped into my

Page opposite: The piano for Steinway. Above: The results of Hurricane Carol.

car, headed for Rhode Island and was arrested for doing 93 mph on the Merritt Parkway but the officer let me go after he heard my story. I pulled into Watch Hill in the late afternoon. The town was under martial law; boats were strewn all over the rocks and roads, with masts leaning into the light wires; the beach club was gone; all the stores were washed out with debris scattered everywhere.

I stopped at the big hotel which seemed to be the center of activity. No one knew much but they were helpful and, after a long while, I located Harriette and Harry at Avard Fuller's house. Apparently Carol had made a 90° turn to the north; the eye passed directly over the center of Watch Hill with winds of over 120 mph. By the time Harriette knew they were in for it, it was impossible to get off the boat as waves were breaking right over the barrier beach. She and Harry let out more scope and ran the engine, taking turns at the tiller to try to hold the bow into the wind. According to the Watch Hill Yacht Club bartender, our boat was pointing right up into the sky at times. The anchor dragged and caught on a thousand-pound mushroom anchor which was all that kept Sea Frolic from going up on the rocks. At one point a large houseboat dragged down on them and did a good job of smashing the bow pulpit and rail. The houseboat occupants wanted to "rescue" Harriette and Harry but she told them she thought she would be better off on Sea Frolic if they could get away from her. They finally wound up on Napatree Beach. Our dinghy was carried away early and then the yacht club dinghy float went by with all the dinghys attached to it, followed shortly afterwards by the beach club, doing several knots. A few hours brought enough calm so that one of the fishermen could come out and take Harriette and Harry off. Avard Fuller came down to the harbor and brought them back to his house where I finally found them. Sea Frolic was the only boat on anchor still afloat. Eroica, Avard's 50-foot yawl, was lying in a field near the Pawcatuck River without a scratch but some distance from the nearest water.

In 1956 WDTA was selected by the Department of Commerce to design the permanent United States Pavilion and exhibit in Moscow. The office was working on the interiors and furniture for the new Air Force Academy as well as the new museum at West Point. My father, Bob Harper and Carl Conrad were all tied up with these major projects so I got the Moscow job, more or less by default. I wasn't especially interested in exhibit work, which had always seemed too impermanent and frivolous to me, but I loved to travel and the design of a large building, something I had never done, was involved.

The first step was to visit Moscow and work out the arrangements, including the selection of a site in Gorky Park, with the Russian Chamber of Commerce. Five of us went on this trip: Harrison T. McClung represented the Department of Commerce, Sydney Jacques the State Department, Bob Sivard the U.S.I.A., Al Stern was a theatrical impresario who had done some good work before for the Commerce Department. I went along as the designer and organizer of the building and exhibit. In October, Moscow was already cold with occasional snow, and everyone wore heavy overcoats and hats outdoors. We stayed in the Internationale Hotel on Red Square, where the rooms were large and old-fashioned with an excess of velvet, brocade and antique plumbing. At that time all U.S.S.R. visits were supervised by the Intourist office which arranged guides, tickets for the various shows in the evening, accommodations and transportation. All the guides—young girls, some quite attractive—spoke good English but varied in their political leanings. Some were obviously skeptical of any statement favorable to capitalist culture, others were intensely curious about life in America. After deciding no one was going to report them, they asked many questions. We ate in various restaurants (*pectopahc*) and in our hotel where the menu was a bit weird but the amount was unlimited including all the caviar one could eat. As a great caviar enthusiast, I had it for breakfast, lunch and dinner and in the evening whenever I could. Fellow guests were interesting. In one corner a group of silent Uzbecs in native costume scowled; at another was a striking brunette with a black, floppy, wide brimmed hat and a black eye patch, straight out of an Eric Ambler spy novel. I was intrigued enough to make some discreet inquiries. She turned out to be an American fashion show organizer from Brooklyn.

Every morning we reported early to the large board room in the Moscow Chamber of Commerce. None of our party could speak Russian but most of the Russians spoke some English and competent interpreters were always present. Everyone was businesslike, anxious to get things done, so the negotiations progressed rapidly. About 10:00 a.m. the meeting would adjourn temporarily and large trays of vodka, cheese, fruit and caviar would appear which everyone would tie into.

Each evening we had free tickets for the theatre or other entertainment. One night we had a fine American dinner at Spaso House with Chip Bohlen, the American ambassador to the Soviet Union. Other nights we visited the circus (terrific), the puppet show (spectacular, with 100 puppets) and Moscow versions of nightclubs. The high percentage of drunks was immediately apparent, a serious national problem. The Moscow subway system was smooth, quiet and efficient; despite all signs being in Cyrillic script, one could find one's way around with no trouble. The spotless stations were ornate in a

Snapshots from Moscow. At work with (from left) Al Stern, Bob Sivard and Sydney Jacques. At an auto show, Dorwin sitting in front of a Zim. Gumm's baronial splendor. Being a tourist.

 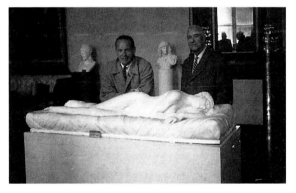

Duke Ellington didn't make it to Moscow—this time—but he held forth often at New York's Basin Street East. Enjoying the jazz, from the left, Craig and Barbara Smyth, Dorwin, Vern Cowper, the Duke, Harriette, Hank and Winnie Gardiner, this photo from the mid-sixties.

Baroque style, all with chandeliers, one of my most lasting impressions of Moscow. Chandeliers were everywhere in hotels, business offices, stores and public buildings. One night the Intourist office informed us that Bolshoi Ballet tickets were impossible due to an unexpected visit by the Premier of Afghanistan. Our State Department representative, Sydney Jacques, swung into action and the decision was reversed. In fact our box was adjacent to the one occupied by Nikoli Bulganin and the Premier of Afghanistan. During one of the long intermissions while we were having refreshments with Chip Bohlen, Bob Sivard remarked that eating caviar with the Ambassador to the USSR in the lobby of the Bolshoi had to be a high point in anyone's career.

Looking for presents to bring home from Moscow was disappointing as there was very little for sale worth buying. In Gumm's, the only department store in town, I was about ready to give up when I spotted a copy of the Polaroid camera at a very good price. I bought one, with some film, and soon found out why it was so cheap. The pictures were almost indistinguishable. This was not the fault of the camera, which was an almost exact copy of the camera that WDTA had designed for Polaroid, the problem was in the film. Actually the camera had a far better lens than the Polaroid in an attempt to make up for the hopeless film quality.

Back in New York I designed the building for the Gorky Park site and got George Rudolph to make a rendering which Commerce liked and approved. My idea was that Duke Ellington and his band would open the U.S. exhibit which I thought would impress the Russians, especially the younger set, more than any amount of State Department or USIA propaganda. I expected resistance from Commerce but Harrison McClung was taken with the idea. The Duke didn't like to fly. He loved trains and composed several pieces based on them. I finally got him on the phone, explained what we were trying to put together and that opening the American exhibit would not only be a big coup for him but would do enormous good for U.S./Soviet relations. He agreed to do it.

This show would have been a monumental success but, sadly, it was not to be. The Soviets chose this particular moment to invade Hungary and, in retaliation, the United States canceled all cultural relations with the USSR indefinitely. Not until fifteen years later, in 1971, would the Duke finally make a Russian tour and the results were just as we expected. His reception everywhere was overwhelming, he received three trunkloads of gifts, and all his concerts, including the one at the 12,000-seat Sports Palace, were sold out.

The Commerce Department felt very badly about the cancellation and in recompense awarded WDTA two alternative exhibits: Vienna, Austria and Zagreb, Yugoslavia. Both included large permanent pavilions as well as the first-year exhibits. After my experience with the abortive Moscow show I was regarded as WDTA's foreign exhibit specialist so took charge of both projects. Zagreb and Vienna are about 250 miles apart by road so a second-hand car, a VW Beetle, was bought to shuttle back and forth as the need arose. Our biggest problem was the time schedule. It was December of 1956 and both shows were to open with exhibits in place by September 10th, 1957. Actually it wasn't until June 1957 that our contract was finalized and we could break ground. The Zagreb building was 100 meters long, plus an office section; the Vienna pavilion, of brick construction, was equally large. We had just three months to build both.

In December 1956 I flew to Europe to meet Paul Medalie of Commerce, who was going to be the manager at Zagreb. I met him in Vienna where we put up at the delightful Sacher Hotel, right next to the Opera. Paul was the perfect manager for an impossible task like this; nothing phased him or disturbed his urbane equanimity and sense of humor. Our key to success in Vienna was Dr. Karl Schwanzer, a prominent Viennese architect, who could not only design buildings and get them up on time, but could also handle every detail of the exhibits including materials, craftsmen and exhibit personnel. Our next stop was Zagreb, the cultural center of Yugoslavia, with the original walled medieval town in the middle of the city. The annual Zagreb Fair (*Velasayam*) was one of the biggest events in Europe, covering hundreds of acres of ground with visitors arriving from all over Europe by air, train, bicycle and foot. Other European countries had permanent exhibit buildings there but ours was to be the first U.S. pavilion. The United States would own the building for ten years after which it would revert to the Yugoslavians. Yugoslavia was a relatively poor nation and supplies of any kind were difficult to obtain. For this reason, Paul and I had decided that our only course was to prefabricate everything in the United States and fly it over.

Among the Zagreb Fair executives we met with was the head of construction, a one-legged World War II hero named Despot. When we explained our plans, Despot made a strong case for allowing his country to do the construction. He said he wanted the chance to show what the Yugoslavians could do when they had to. After a long talk back at the hotel, Paul and I agreed to let them do it. This was rather a scary decision but, as it turned out, the right one. The Yugoslavians not only made every deadline but the mutual bond of respect between our people and theirs was the best kind of diplomatic relations.

On Christmas Day we had a big dinner with various USIA and Consulate people in Zagreb. This was my first Christmas away from home so I wrote out a Merry Christmas message to Harriette and took it down to the cable office. There I was informed the language was not permissible in this non-Christian Communist land so I had to change it to say "Happy Jack Frost Day" much to the delight of the rest of the U.S. group. Actually most Yugoslavians didn't take the Communist doctrine very seriously. Anti-Communist jokes were popular. The USIA and State Department representatives I met were dedicated, intelligent people and consequently there was a great respect and admiration for our country.

Because I had brought along my ski boots, I was able to get in a little practice in a Zagreb park on some borrowed skis. On the way home I stopped in Davos, Switzerland. This was ahead of the regular season so I got a good room in the best hotel and rented a pair of fiberglass skis from the son of a local ski shop owner. The only other passenger in my first ride up on the funicular was one of the instructors who invited me to join him on the fourteen-mile Parsenn run. Conditions were perfect, I turned wherever he did and we stopped at a little chalet for a hot chocolate and a gluwein halfway down. After a meal in Closters we took the train back to Davos. On January 7th of 1957 I flew home to Alpine, New Jersey for a respite.

The working group for Zagreb and Vienna included George Gardner, Jack Field, Bob Blood and Boris Nepo who drew up the plans for Zagreb. Boris was a White Russian and the fastest architectural designer/draftsman I've ever known. Fluent in Russian and English, he was able to pick up Serbo-Croatian rapidly and could converse with the Yugoslavians easily. The Vienna building was a simple, modern structure with very striking exterior graphics on the Vienna fair grounds where the giant ferris wheel from the 1900 Paris World's Fair is located. George Gardner did the design of the building, the graphics and the exhibits in New York and Dr. Karl Schwanzer helped George to oversee the construction in Vienna.

The Zagreb pavilion was on two levels, the ground floor and a large balcony, but the front of the building consisted entirely of large louvers, each one by five meters in size. This was my basic design. The

Fair was held in September when the weather was usually perfect and the idea was for a building with the look of an open showcase so visitors could see the inside as they approached and the exhibit could run right out onto the courtyard in front of the building without interruption. The louvers were connected together and could be closed in case of storms. The exhibit part, 100 meters long, had a Y-shaped cross section; the main structural members were a series of eleven massive branched steel I-beam sections. These were fabricated in the Yugoslav steel mill in Maribor, about eighty miles from Zagreb near the Austrian border; the mill operated twenty-four hours a day in shifts to meet our schedule.

In April I made a second trip to Vienna and Zagreb to meet with Paul Medalie and finalize the arrangements with Schwanzer and Despot. On the way I stopped off in Stuttgart to visit Klaus Scheufelen. This was my first visit to his home town and I was greatly impressed by his paper factory which had just installed a 145-inch-wide continuous coated paper machine which ran at some incredible rate of speed. I also had a chance to visit Klaus' summer place on the Starnbergersee, where he kept and raced a Dragon class sailboat. Knowing my interest in automobile racing, Klaus threw a cocktail party for me that included several famous drivers. One was his next-door neighbor Ernst Henne, who for years held the world's motorcycle speed record on BMWs and was then the Mercedes distributor for Bavaria. Rudi and Alice Caracciola drove up from their home in Switzerland. The world's champion automobile racer before the war, Rudi was known as *"Der Regenmeister"* for his skill on wet tracks. Alice was one of those people who can switch from perfect English to perfect French, Italian or German as required, so she took over the job of filling me in whenever my meager German failed me.

The story told that night that I remember best concerned Caracciola and crew returning from some race in a large Mercedes with *Der Regenmeister* driving. In a typical small German town with narrow streets, a bicyclist came wobbling out in their path. It was impossible to turn in either direction or to stop; the rider was hit amidships and both he and bike were tossed up in the air to come down in back of the car. Unconscious, the bicyclist was bleeding copiously from nose and mouth. While one of the crew ran for a doctor, the others tried to do the best they could for the unfortunate victim, fearing the worst. As the doctor came up on the run, the cyclist came out of his coma, rose and started brushing himself off in front of the horror-stricken crew. His handlebar basket had been filled with freshly picked raspberries and he had landed head first in it after his trajectory over the car. Aside from a bump on the head and a bent bicycle, he was as good as ever.

Around June 10th the contract was finally settled. We broke ground for both buildings and, thanks to Boris and George, everything progressed on schedule. I designed the louvers for Zagreb in New York, using two sheets of corrugated aluminum roofing material riveted together with aluminum end plates, and stainless steel pivots and connecting rods. None of this material was available in Yugoslavia so we shipped all the raw material over and Despot made arrangements to have the louvers made to my

Oberlinnigen and the Scheufelen paper factory, Klaus (right) chatting with his brother Karl.

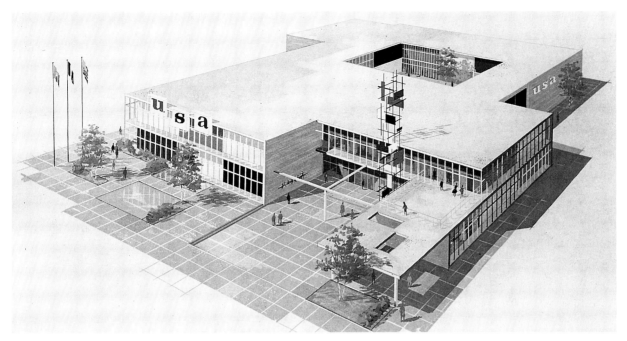

Above: Models of the permanent U.S. pavilions for Vienna (above) and Zagreb (below).

drawings in thirteen different Zagreb machine shops. Everything went together perfectly. Our people and the Yugoslav construction crew worked side by side, communicating in sign language and ending up with "*dobro*" which signifies "okay, let's go" in Serbo-Croatian. The exhibit for Vienna was a mixture of American products and machine tools and proved very popular. For Zagreb the main theme was an exact reproduction of a working supermarket where Fair visitors could see how Americans shopped for food. The Food Chain Association did a fantastic job of organizing and installing its exhibit, leaving nothing to chance and shipping everything over, including such odd items as bales of rags.

In September, just before the opening, I made my final visit to Vienna and Zagreb. There are always a lot of last-minute problems to clean up in those exhibits but everything was under control, and we couldn't have done any better if we had had an extra year to get ready.

Paul Medalie was an avid upland game shot. Because he knew I was also interested, he organized a pheasant shoot near a country village named Dugo Selo (which translates to "long town" in Serbo-Croatian) about twenty kilometers east of Zagreb. This was a feudal European-type shoot with about a dozen beaters under the direction of a wiry five-foot peasant with a loud voice and a pack of dogs. There

Below: A Yugoslav worker on the job in native dress. Page opposite: Zagreb's layout; Dorwin on forklift, with Al Stern (left) and two Yugoslavs looking on. Paul Medalie (with glasses) escorting Marshal Tito and his wife at the fair. Photos: George S. Gardner.

were about six "guns" including the Communist Secretary of Zagreb, the young Mayor of Dugo Selo, Paul and myself. He and I were equipped with beautiful engraved over and under 12-gauge shotguns made in East Germany.

The European shoot is a ritualized affair; the "guns" are stationed outside a copse while the beaters and their dogs go in from the other end, advancing in and driving out the game ahead of them. All kinds of game come out including large hare, a few small wild pigs and deer, but we were only interested in pheasant. When the first pheasant came out, there was yell of "*coca*" (like in Coca-Cola) which Paul naturally assumed was Serbo-Croatian for cock pheasant so he promptly knocked it down. Unfortunately "*coca*" means "hen" in Serbo-Croatian which is illegal game just as it is here in the States. Everyone had a good laugh after the mistake was explained. Dinner at the end of the day was at a long banquet table, with beaters and all, where we ate our trophies. Then the Yugloslavians sang Serbo-Croatian hunting songs in beautiful harmony, led by the mayor of Dugo Selo, with the little head beater contributing a powerful bass. At one point the mayor asked me if I would sing our national anthem. Anyone who has ever tried to sing the *Star Spangled Banner* solo, without accompaniment, and with an imperfect knowledge of the words, will sympathize. Fortunately I started in a low enough key so that I could manage the high part and with a certain amount of faking of the words, I managed to get through it.

On September 7th Marshal Tito walked down the aisle of the Zagreb exhibit and cut the ribbon. The supermarket show was characterized in *This Week* magazine as a "public relations knockout." Zagreb radio presented a flattering program in praise of the building as a happy cooperative effort between Yugoslavia and the U.S.A. Incredibly the total budget for the building was $235,000 which included the large enclosed and air-conditioned two-story office building at one end.

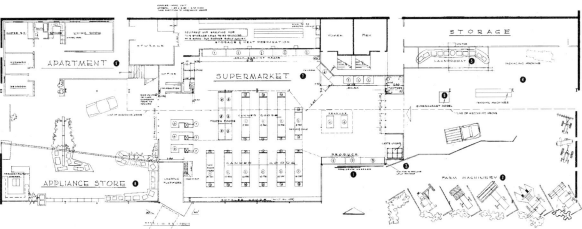

129

Fairgoers were enthralled with "Supermarket U.S.A." About thirty Yugoslavs staffed the exhibit each day. University of Zagreb students worked checkout. Photos: George S. Gardner.

My fourth European trip that year was to supervise the installation of a large model of Idlewild Airport (now JFK) which the WDTA model shop had built for the New York Port Authority and which was to be installed in the International Building at the soon-to-be-opened Brussels World Fair. As usual, everyone else was tied up so I was elected to go. The International Building was designed by a Belgian architect named Montois and, with its huge symmetrical airfoil section roof, was quite interesting. After supervising the installation, I had dinner with Montois and his family, none of whom spoke any English whatsoever so I had a chance to brush up on my French.

With my European commuting and other work, I didn't have much time for sailing and the hurricane damage to Sea Frolic was serious enough so that Harriette and I decided to sell her. For several years we were "on the beach" with no boat so I bought a new, bright red Austin-Healey 100/4 sports car that I ran in a couple of rallies and a hill climb on Mt. Equinox.

My next car was a Porsche 1500 Speedster, the first of six Porsches I owned over the next ten years. I also joined the Sports Car Club of America and got a temporary permit to enter the SCCA drivers school, held mostly at the Lime Rock and Thompson race circuits in Connecticut. This brought home the amount of experience and ability required to become even a reasonably good competition driver. An interesting phenomenon to me was the fact that one's reaction time, which I had always thought was inherent and immutable, can be trained to operate faster. On the way home after one of those driving sessions, the normal traffic seemed to be moving in slow motion. Even today, many years later, my reactions are still much faster than those of most young people.

The Austin Healey, model 100/4. Left: Dorwin driving to second place at Mt. Equinox. Center left: At Seven Lakes, where Dorwin took first overall. Center right: Harry in the Austin Healey, his cousin Allison in Uncle Lewis' Baby Bugatti. Bottom: At Lewis' home. The Porsche 1500 Speedster that replaced the Austin Healey.

Chapter Twelve

1958 – 1959

Clyde Cowan had the ability to explain recondite theories in understandable terms ...

Bugatti Type 37. With Ben Stansbury (above) and with Amherst Villiers' wife Nita (page opposite). Because Ettore Bugatti was a devoted equestrian and his cars were pur sang (thoroughbreds, in the vernacular), owners routinely sought out a picture with a horse. Dorwin's T.37 with Lesley Turner on Domino.

With no boat to occupy our time, I spent most weekends and vacation periods in cars. An opportunity came up to buy a Type 37 Bugatti, the same model I had first seen and admired years before at Gilhooly's. A four-cylinder 1500 cc model built in 1929 on a Grand Prix chassis, the Type 37 had an open, staggered two-seater body covered with lots of little louvers, straps to hold down the hood, a tiny wind screen for the driver only and a boat tail in the back. The brake and gearshift levers were mounted outside the body on the driver's side, with the spare wheel and tire strapped alongside on the left. The Bugatti horseshoe radiator was in front. The squared-off motor, engine turned on the outside, looked like a piece of sculpture. The car had been driven by a long series of owners in England and America, including Tony Hogg who later became editor of *Road and Track*. It was a little rough around the edges but I had a good garage/shop in our house in Alpine and was able to iron out the worst spots. A 100% restoration didn't interest me; I felt the car was more interesting in its used form, looking like it had just been cleaned up after finishing the Mille Miglia. The running gear, springs and axles were all polished, according to standard Bugatti practice.

Like all Bugattistes, I joined the Bugatti Owners Club in England. In addition to a tri-monthly publication called *Bugantics*, the club published a book, updated periodically, which listed every known Bugatti in the world with particulars on the owners and the cars and notes on any non-authentic parts. I know of no other make of car that has created such a loyal, enthusiastic following. Despite the Bugatti's many faults, Bugattistes remain fierce protagonists. I drove my Type 37 to occasional meets and concours, and even commuted to work in

Manhattan once in a while, after making sure of the weather forecast. The car had no starter so had to be hand cranked which could be embarrassing if it stalled in traffic.

In 1958 the Sports Car Club of America held a veteran car meeting at the recently completed Bridgehampton race track on Long Island which replaced the former dangerous road course through the town streets. The track, with a number of turns and elevation changes, together with some fairly long straights, was quite interesting to drive. The climax of the meeting was an "exhibition run," actually a thinly-disguised race of about ten laps of the course. My Bugatti was among the dozen entrants. A French Delage racer won; the driver was Briggs Cunningham, who successfully defended the America's Cup that same year. Second place was taken by a 3-liter Type 43 Bugatti driven by Mark Donohue who later became one of America's best race drivers. I was third in the Type 37 and have a small cup to commemorate the event.

My appetite was whetted. After obtaining an SCCA competition permit, I entered races in Connecticut, Bridgehampton, etc. and, after a year or so, traded the 1500 Porsche for a 1600S Speedster which was faster and better for racing. Harry and Harriette would usually come along to the races.

The year after the Zagreb and Vienna fairs I was given a contract by the Atomic Energy Commission to design and oversee construction of the United States exhibit at the second international "Atoms for Peace Exhibition" in the Palais des Nations in Geneva. The USSR had just taken the lead in the space race with the successful launching of Sputnik. To counteract the Russian coup, a substantial sum was allocated to emphasize U.S. superiority in nuclear research and technical applications of atomic energy. In the fifties nuclear power was still thought of as the ultimate solution for clean, inexpensive power generation. The budget for the ten-day show was several million dollars; all the facilities and many of the top scientists of the AEC were put at our disposal.

Competition among the many national participants at Geneva was intense. Our principal adversary, the U.S.S.R., was predictably building its exhibit around the Sputnik success. The A.E.C. wanted to demonstrate the rather esoteric scientific advances in nuclear physics with graphic and three-dimensional displays which would be understood and appreciated by all attendees, regardless of their level of scientific knowledge.

Our principle contact was Dr. Clyde Cowan, co-discoverer of neutrino and first to observe the trace of the tiny particle. The combination of higher mathematics, imagination and vast learning necessary to postulate the theory that such a particle exists, and then to devise a way to trace and prove its existence, is beyond ordinary comprehension. But Clyde Cowan had that rare ability of being able to explain his recondite theories in simple and understandable terms. His clear thinking was just what we laymen needed in order to devise dramatic

Porsche 1600S Speedster, dressed for racing and skiing (the non-invasive approach). As No. 35, on the way to Thompson, No. 28 arriving at Lime Rock, and No. 47 at the start and in full cry at Bridgehampton.

The "Atoms for Peace" exhibit designed for the U.S. Atomic Energy Commission, 1958.

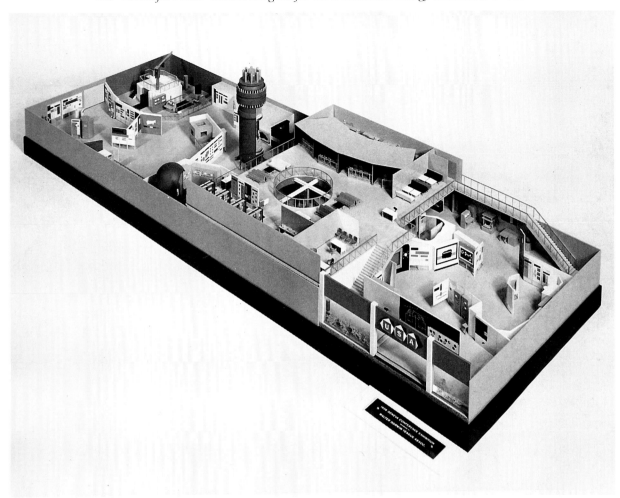

and understandable expositions of abstruse theories and developments.

The 10,000-square-foot U.S. exhibit included a number of interesting and dramatic displays which could be appreciated and understood by third-world technicians as well as experts in the particular field. One effective feature was a sound film library covering hundreds of subjects related to atomic physics, with four screening areas where visitors could request a film on their choice of subject in English, French, Spanish or Russian. Another dramatic display was an actual low-level working atomic reactor that started on the first day of the symposium as nothing but a hole in the ground, was constructed during the ten days of the show and, on the next to last day, was producing power. (The procedure had been fully rehearsed in the U.S. to make sure it could be done.) Visitors returned each day to read the posted progress reports and watch the construction.

George Gardner was in charge of the project, joined by the same Vienna and Zagreb exhibit group which had evolved into an efficient and effective team. After the design was finalized, an elaborate scale model of the entire U.S. exhibit was built in the WDTA model shop so A.E.C. officials could visualize it. Later on the model also helped in the actual construction. In April of '58 I flew to Geneva to make the preliminary arrangements and on the way made a side trip to Molsheim to visit the old Bugatti factory and pick up parts I needed for my car. In July, when the actual construction started, George Gardner, Bob Blood and Jack Field moved to Switzerland for the duration and rented one floor of a villa near the Palais des Nations. When I next went over in August, just before the exhibit was to open, Harriette and Harry came along. We took an apartment in a little hotel in Nyon, a few miles north of the Palais, with a large outdoor balcony overlooking Lake Leman where we ate breakfast in the morning and read about the previous day's motor scooter accidents in the local paper. A pleasant little

Atoms for Peace exhibit; atomic reactor model in the background. Above: Paying a visit, from the left: Clyde Cowan, Captain Edward Gardner (head of Atoms for Peace), Dag Hammerskjold and AEC chairman Robert Straus. Below: The Bubble Chamber, or what happens to a tank of water when neutrinos descend from space and hit other particles. Photos: George S. Gardner.

Left: The Teagues' hotel in Nyon, picnic at Prescott, watching the races. Above and below: The nicely deserted cove in Majorca.

town, Nyon was celebrating its 2,000th anniversary which I thought at first was a misprint. (Nyon was established as a rest camp for Julius Caesar's officers.) The French owner of the villa where the WDTA team was staying fancied himself as a gourmet, so by the time we arrived our people knew all the really fine restaurants in the area. This was valuable knowledge as many of the best places had small obscure signs which one would ordinarily never notice. At noon everyone repaired to the nearby beach to have lunch, swim and water ski.

Our exhibit was the hit of the show. After it opened, Harriette, Harry and I flew to Majorca where we had a suite in the best hotel in Palma for the equivalent of six dollars a day. We hired a funny little three-wheeled car made by Voisin which barely held the three of us plus our scuba gear. Our regulators and masks were brought from home; tanks and weight belts were rented in Palma. This section of the Mediterranean was crystal clear and ideal for scuba diving. On our first day, we got off the main highway and managed to get thoroughly lost on country dirt roads, arriving by chance at one of the most beautiful deserted coves I've ever seen. This became the base of operations for the rest of our stay. We never saw another person there except for an attractive young English couple who also ran into it by chance. Years later I recommended the cove to my friend Tip Sempliner, who was going to live in Majorca for a year. He found it but, sorry to say it was completely built up with condominiums, the fate of many of the most attractive sections of the Spanish coast and islands.

From Majorca we flew to London, checked into the Savoy and we did a lot of sight-seeing with the Fishers as guides. Now fourteen, Harry was gregarious and happy so strangers responded in kind. He enjoyed London tremendously. A Bugatti Owners Club hill climb was upcoming at Prescott, about 100 miles from London through pretty country so I hired a car for the journey and asked the Savoy to make us some sandwiches to take along. Our rented Humber Hawk was all packed, but no sandwiches. I became anxious. Finally a couple of porters came out carrying a huge fitted wicker trunk, just barely able to fit into the luggage boot; our light lunch turned out to be a gourmet meal with soup, wine, chicken in aspic, various kinds of meats, salads and desserts, heavy silver tableware for three, silver condiment pots, stiff linen napkins and a large linen tablecloth to spread out on the grass.

After we got home, Harry and I began talking about the *"pedalos"* we had rented in Geneva. These were two-seated pontoon boats with a set of foot pedals driving a paddle wheel and a rudder. I started putting some rough ideas on paper for a powered version. To my knowledge there were no powered pontoon boats in 1958. Our idea was to use surplus materials and do it as cheaply as possible but make it seaworthy enough to cruise on the Hudson River. One of my old racing outboards would be used for power. In the surplus war materials section of the Sunday *New York Times* I found some new surplus F86 aluminum wing tanks fifteen feet long and two feet in diameter for $50.00, which had undoubtedly cost the taxpapers several thousand a piece. These had ample displacement and rugged mounting brackets, designed to take the shocks and high "G" forces of fighter plane maneuvers and landings, so would be plenty rugged for a boat. I designed a subframe of extruded aluminum I-beam sections and bought about $150.00 worth of stainless bolts, nuts, washers, bronze lag screws, monel wood screws and boat nails so that the whole craft would be corrosion resistant. The main frame was 2x6-inch pine with four pieces of 4x8-foot plywood to form the 8x16-foot deck. The varnish was also war surplus and I begged some old fire hose from the Nyack Fire Department to make bumper strips down each side.

The motor mount was on a hinged cut-out section aft that could be adjusted up and down to conform to the loading. On the spacious deck I installed a cheap helmsman's seat, a tool box, a gas tank and a large tin box for gear that I liberated from

Building the Section 5.

under Harriette's bed. The wing tanks were fine as buoyant hulls but couldn't stand concentrated loading on land so I designed a plywood Roman chariot which fit between the floats and ran on a pair of aircraft tail wheels, also war surplus. We towed the whole rig behind Harriette's VW Beetle, a strange sight on the highway as the boat dwarfed the car. We named our craft "Section 5"—after the Sunday *Times* section in which war surplus was advertised—and had an official christening, champagne and all, at Cornetta's launching ramp in Piermont, New York. The Section 5 was a big success from the beginning. Our first cruise, to the George Washington Bridge, went so well that we decided to continue to the Battery and circumnavigate Manhattan. Low on fuel on the East River we pulled in at 23rd Street and took our fuel tank in a couple of blocks to a gas station for refilling. Continuing on up through the Harlem River, we discovered that Section 5 had one unique attribute; she could pass under any closed bridge with a vertical clearance of over two feet. On a later cruise we even passed under the New York Central tracks at the entrance to the Van Cortland Manor restoration at Ossining, though we had to remove the cap from the fuel tank and lie flat to get sufficient clearance. That summer I had the Bugatti in a veteran car meet on Randall's Island and got to talking with Michael James of the *Times* about Section 5. He came out to Alpine the following week with a photographer and wrote an article on our craft that appeared on the front page of the second section of the *Times*.

Exhibit assignments continued to come my way, including a temporary building in Amsterdam for the American Horticultural Society exhibit in the "Floriade." Another job was the new entrance complex for Longwood Gardens, the large DuPont public gardens and conservatories at Kennett Square in Pennsylvania, which entailed the design of a new highway intersection, a parking lot for 1,500 cars which had to be invisible from anywhere in the Gardens, and an Information and Reception Center with snack bar and offices. A quite complicated project was the design of several mobile, expandable trailers for the Foreign Agricultural Service, part of the Department of Agriculture, which found itself with a large quantity of so-called "blocked pesetas," funds that had to be spent in Spain. The plan was to build the three-trailer exhibit in Barcelona and then ship it through France to Italy for an exhibit. The logistics of this international transaction were horrendous. After everything was finished, Bob Blood—who had designed and superintended the construction—found that the prime contractor had de-camped with all of the funds and none of the sub-contractors had been paid. Bob arranged for a lawyer to represent the sub-contractors and I believe they all got their money eventually.

Page opposite: Dorwin, with Harriette and Harry, enjoying a cruise on the Hudson with Section 5, possibly the world's first motorized pontoon boat. The novelty of "powering a catamaran," as it was termed then, attracted the press— and stories followed in the New York Times *and* Popular Boating. *Right: Longwood Gardens, the entrance below.*

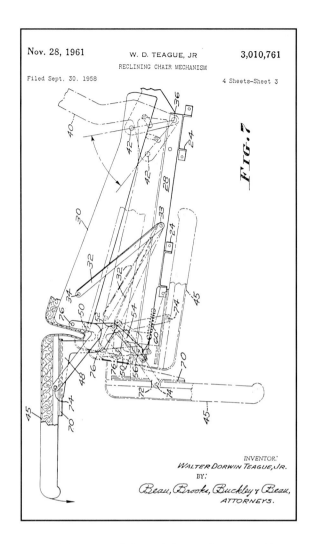

In addition to all this exhibit work, I was still responsible for the Navy inspection program which had grown in scope but, from time to time, I had the chance for product design and development which was still my first love. One of the most interesting assignments was the development of a new type of dental chair for the Ritter Company, the foremost American dental equipment supplier.

By this time I had hired Maurice Vauclair, my chief draftsman/designer at Bendix, to work for WDTA and with Ben Stansbury and Dave Deland, another clever designer, we made a strong development team. Ritter Shumway, Ritter's president and our contact was very able and pleasant to work with. And interesting: he took up figure skating in middle age and by 1953 had won the national veterans' pairs competition.

The chair was originally the concept of a California dentist named Golden who figured out there was a certain contour which would provide a comfortable reclining seat for any person, regardless of size, sex or build. By reclining the patient, the dentist could work sitting down rather than standing, which had caused job-related health problems. A plywood mock-up of his ideal contour was about all we had to start with. Golden had received a patent on his idea and had sold Ritter Shumway on adopting it for his line.

We did all the engineering on this chair including production drawings for all parts except those in the original hydraulic system, and we specified all materials. Asked by Ritter to recommend a supplier for the fiberglass seat, I arranged for the production to be given to the Anchorage in Warren, Rhode Island, builders of the famous Dyer dinghies which were then and still are the highest quality small fiberglass craft in the world.

The outcome of this project was all one could hope for. Our design was accepted without modification, went into production and was highly successful on the market. Many of these dentist chairs are still in use. The design is dateless and still looks beautiful and right. The Industrial Designers Institute, as the American Society of Industrial

Recliner chair patent, Ritter dental chair patent and later public relations release about it.

This prize-winning design by Dorwin Teague for the Ritter Company included all experimental and production engineering as well as appearance design.

Designers was called then, awarded Ben, Dave and me the prize for the best product design of the year in any category. Italian designers paid us the ultimate compliment by closely copying our design and selling it here at a lower price.

Cruising on the Section 5 was fun but no substitute for sailing. At the 1959 boat show the newly organized Pearson Company of Bristol, Rhode Island showed the prototype of a 28-foot fiberglass cruising auxiliary called the Triton, designed by Carl

*Below: The prize-winning chair and Ritter Shumway, Dorwin, Ben Stansbury, Dave Deland.
Page opposite: Olé, Triton No. 8, the first fiberglass cruising sailboat on the Hudson River.*

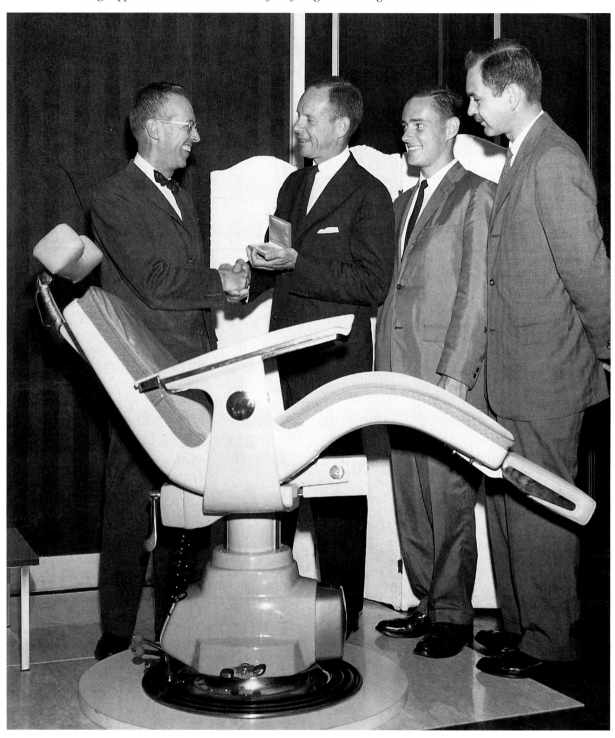

Alberg of Marblehead. Listed at $7,995 without sails, the Triton became the first large production fiberglass cruising boat. Sales eventually reached over 600 units; a well-maintained example sells for $15,000 today. Offered in sloop or yawl rig, auxiliary power was a Universal Atomic Four. For me the Triton was love at first sight. Naively I assumed that, with the new fiberglass construction, the days of leaking decks and hatches were over. With no frames or ceiling, the amount of interior room was more than in our previous 34-foot wooden boat and her 3/4 rig made her easy to handle. Her lines were beautiful. In spite of the fact that no Triton had ever been in the water, I put down a deposit to become owner of Olé, Triton number eight. As one might expect, the brand-new company had a lot of teething problems; the exterior ice box access was a disaster, the sprayed-on interior overhead treatment developed rampant dandruff soon after delivery, hatches, ports and chain plates all leaked profusely. Delivery was late, of course, but we finally got under way.

Aside from a rather strong weather helm, Olé was a delight to sail. During our first summer we cruised to many ports in Nantucket Sound, Vineyard Sound and Gardiners Bay including a visit to Gardiners Island with Hank and Winnie Gardiner which Hank arranged with his cousin Robert.

Chapter Thirteen

1960 – 1962

There is no way to tell a can of beans from a can of apple sauce ...

In the late winter of 1959-60, I turned an ankle skiing which kept me off skis for the rest of the season. To take up the slack, I picked up a copy of *Mixter*, the class work on celestial navigation, and a surplus bubble sextant. Before very long I was able to pinpoint my location most of the time with reasonable accuracy so I bought a proper sextant, and the necessary tables and almanacs, and began to practice sights from various shore locations where I could see the horizon line. Taking sights and working out one's position on land is all very well, but a long way from the real thing at sea on a small boat bouncing around in all directions. There is only one way to experience this; one must go out to sea in a boat and do it. We had an extremely seaworthy boat and, in spite of being only twenty-eight feet long, I believed Olé could take anything that might come along. From that moment on all my spare time (and spare cash) was concentrated on a sea voyage.

The choice of destination was easy. None of us had ever been to Bermuda, which was only some 600 miles away and could be reached easily in our normal vacation period. Furthermore, the bi-annual Bermuda race would be held the following June and, if we scheduled our voyage a week or so before the race, there would be a lot of boats coming along in case we were seriously disabled. The fact that Bermuda is a relatively tiny island, far from land in any direction, made all the more of a challenge to find it. Offsetting this was the large quantity of data available on the trip as a result of the Newport-Bermuda races. From this information, it was probable we would run into some dusty weather and set out to prepare for this. As a starting point, using the rules of Cruising Club of America, we equaled or exceeded them in every respect. I ran a second sail track up the mast as far as the spreaders and had Ratsey & Lapthorn make up a tiny storm trysail and storm jib. In addition to ordering a complete spare tiller assembly, I made up a titanium-lined hole in the rudder to reeve steering lines through in case both tillers failed. Our local shoemaker sewed up four safety harnesses out of surplus nylon belts and hooks. Enough food and water for several weeks would be aboard in case we got really lost and a single burner Sea Swing stove was installed, for which a local distributor gave us a carton of "Handy Fuel" canned heat. The labels of all cans were removed and the cans marked with nail polish; many voyagers have learned too late that paper labels come off in the bilge and clog the pumps—and make each meal a lottery as there is no way to tell a can of beans from a can of apple sauce. In addition to the sextant we carried an Accutron watch, a Channel Master radio for time signals (exact time is essential for determining longitude), a Heathkit radio direction finder and a surplus Gibson Girl transmitter for emergencies. The crew for the trip included Harriette, Harry, who at age fifteen was large, strong and agile, and Ben Stansbury who had no sailing experience but who I suspected would be a good sailing companion. This turned out to be the case.

Our departure from New York was set for June 10th, 1960. I look back on that spring, when we were concentrating on the one goal, as among the most exciting and satisfying periods of my life. The fact that none of us had ever been out of sight of land in a small boat, and had never been to Bermuda, made it an exciting adventure, not necessarily dangerous but challenging and thrilling, a perfect antidote to humdrum everyday existence. I've skippered my boat in a number of ocean races since then, including four Bermuda races, but none of these equaled the satisfaction and feeling of accomplishment that this first little voyage brought about. I think the person who never seeks adventure misses a great deal out of life.

As for the trip itself, we got the predicted mixture of good and bad weather including one storm which blew us ninety miles east of the rhumb line under bare poles with anchor rodes trailing astern to slow us down. We got one knockdown and narrowly missed losing our mast. Everyone on board got seasick at one time or another. But the first sight of the pink and white coral houses in the warm early morning sun, seven days after we started, made up for all the work of preparation and hardship. Despite being busy getting ready for the finish of the race, the Bermudians outdid themselves in hospitality. We were given guest cards to the Royal Bermuda Yacht Club; Llew and Helen Gibbons gave us one of their moorings at Point Shares with a nearby guest cottage for showers. Ben and I flew back to New York after a short stay on the island and, Olé was shipped back

Olé, 1960, right from the top: navigator's shelf, view from main cabin (note navy surplus chart lights), galley with sea-swing stove.

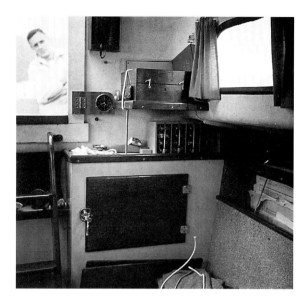

to New York on the deck of the *Queen of Bermuda* with Harriette and Harry in one of her cabins. Soon after my return to New York, I wrote an article about our Bermuda trip which was published in *Yachting*, the first of many I would do for boating magazines.

My mother died on the 10th of September in 1960. Her ashes were buried in the little family graveyard in the field below the Brick House that she loved so much. On the 3rd of December that same year my father died of emphysema in the hospital in Flemington, New Jersey. This left Bob Harper as head of WDTA, me as number two partner. For the

The Bermuda trip, 1960. Leaving from New York, Dorwin still dressed for work. Harriette at rest. Ben Stansbury "on the ropes." Harry hoisting the British flag on approaching Bermuda harbor. Hoisting Olé for the trip home.

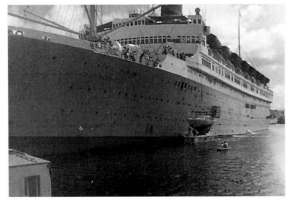

time being very little changed in the running of WDTA since most of our business was in long-term accounts like the Boeing work run by Frank del Guidice and the Navy contracts which Sy Gollin and I directed.

In 1961 I was retained by the Virginia Civil War Commission to design the Civil War Centennial Information Center building in Richmond, which later became the library for the University of Virginia. A circular domed building, reached by ramps to the second floor, it enclosed dioramas and exhibits which we designed. We also became movie producers for this project and retained a New York firm to do a documentary to illustrate Virginia's part in the Civil War, using a combination of live action with flashes of Matthew Brady photographs and artwork. Joe Cotten, who was born in Virginia and an old friend of our family, did the narration very effectively. The only glitch was the background music chosen by the director for the final theme: "The Battle Hymn of the Republic." The melody is stirring and effective but, of course, the lyrics, which were not used, are violently anti-Confederacy. This choice infuriated the Virginia Commission members. We apologized profusely and quickly changed to "Dixie" but I'm sure there are members of the Commission to this day who are convinced we acted maliciously.

One of our design accounts was a new boat for the Pearson Corporation that came about as a result of my purchase of the Triton. After our Bermuda trip, which received some publicity, we began to race Olé in the Eastern Long Island Sound series and a few distance races. I became annoyed at the extreme weather helm and finally took the drastic step of cutting the main boom down from fourteen to twelve feet and had the mainsail re-cut accordingly. We began to beat other Tritons consistently. Our friend Tom Harrison and some of our other competitors followed our example. Finally the Pearson Corporation made this a production modification although, as I remember, a bit more conservatively by cutting back the foot of the sail by only 1-1/2 feet. A more serious inherent fault began to show up as several Tritons had mast failures during hard blows. Olé was leading a race off Larchmont by a good margin, when our mast suddenly went over the side. By this time the insurance companies were getting a bit fretful and the Pearsons (Everett and Clint, who are cousins) decided something had to be done so they hired me as a consultant to try to find out what was going on.

I inspected five or six failed masts in the Long Island Sound and New England areas and discovered all the failures were identical. The mast, stepped on deck in the Tritons, had all failed by bending aft at the spreader line. This brought to mind the knock-

Virginia Civil War Centennial Center (now the University Library). Photo: Virginia Chamber of Commerce.

The 28-foot cruiser for Pearson. The gangplank system for the Port Authority of Puerto Rico.

down on our Bermuda trip, after which both spreaders were angled forward. I couldn't figure this out for a while but finally the answer came to me. Carl Alberg had designed the Triton with a single lower shroud, a legitimate system if done right. But there was a bulkhead in the Triton a foot aft of the mast line; the Pearsons assumed the chainplates for the lower shrouds would be much more secure if attached to this bulkhead rather in line with the mast, as called for in Alberg's design. What this did, however, was to produce a strong after-force component on the mast right at the spreaders, a weak point anyway because of the hole for attaching the spreaders. We were lucky that our mast hadn't failed on the way to Bermuda. We designed a "recall" kit for the existing Tritons, while the Pearsons corrected the problem in later ones.

With some confidence in my design ability now, the Pearsons asked me to design the new 28-foot power cruiser they wanted to add to their line. I agreed but wanted someone with more marine architectural experience to work with me on it. The British-based Amateur Yacht Research Society, of which I was a member, was concerned with all sorts of wild ideas and experiments with far-out sailing concepts. Its American section was headed up by a young, enthusiastic marine architect from the East Indies named Walter Bloemhard. Walter was rather a genius and, like all genius types I've known, had certain individualistic personality traits. But I thought he would make an interesting addition to the WDTA staff so I hired him. We collaborated on the cruiser with Walter doing most of the hull calculations: the boat turned out well with excellent performance and attractive accommodations. It became the best seller of the Pearson power boat line.

Walter fitted in well as a member of my group. One of our design clients, the newly organized Enjay Corporation, wanted some ideas for publicizing their new polyurethane sheet material which, when bent sharply or folded, actually became stronger as the molecules lined up. This seemed a natural for Walter and two of his ideas were shown in company ads in various technical magazines. Soon after the folding boat ad appeared, I got a panic-stricken letter from a man in California who had just come out on the market with a very similar version of a folding dinghy. I told him that we had been well paid for our concept and wished him all success with his boat.

A new responsibility was added to my work at WDTA when I took charge of a project for Puerto Rico. "Fomento" was the government's new venture to give impetus to industry and the tourist trade by improving Puerto Rico's image and facilities, particularly in the area of design. When the head of the Fomento group visited our New York office, we started giving him our usual sales pitch but he cut us short, said they had thoroughly investigated all the prominent industrial designers and had already decided on WDTA. All the expenses involved in opening an office in San Juan would be paid, and Fomento would underwrite the entire operation, salaries and all, for a year. For the second year, we would be paid only a percentage of any losses and after three years, as I remember, we would be on our own. This was an irrestible inducement so we accepted and I was put in charge. Laurie Williams was made head of the office there, succeeded in 1961 by Russell Tonkiss who spoke fluent Spanish with a Castilian or Puerto Rican accent as required. This was pleasant duty for me as it called for periodic visits to San Juan which is only a short hop from the Virgin and Leeward Islands of the West Indies. I introduced Russ to skin and scuba diving at which he became adept. Russ was basically a graphic designer, the main need in Puerto Rico then.

One job I worked on personally was the design of a new cruise ship pier in San Juan. At that time all passenger ship piers had huge superstructures running the length of the pier whose purpose was to sling the great heavy gang planks from pier to ship for disembarking passengers and baggage. It seemed to me that there must be a better way to do it than this unsightly and expensive system so I decided to rethink the whole problem. First I had the WACO Company, experts in light metal structures, design an aluminum gangplank that was adequately strong but light enough so it could be jockied into position on any size ship by one man with a specially equipped fork lift truck. There was a passenger version and a

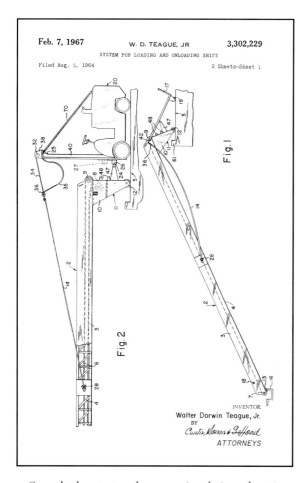

Gangplank patent and a promotional piece about it.

conveyor version for baggage. I received patent number 3,302,229 on the idea which did away with all superstructure, was less expensive to make and required less manpower to manipulate. The original prototypes and several duplicates are still in use on the two major San Juan cruise ship piers.

As a result of our Puerto Rico work I began to visit the Virgin Islands where I usually stayed at the Club Comanche in St. Croix. Bill Miller and Dick Newick, who were bitter enemies, ran competing day trips to Buck Island. One of Dick's captains was the legendary Bomba, the best natural sailor I've ever known. He never wore any shoes and was immensely strong; a friend of mine once gave him a case of beer and Bomba walked off, swinging it in one hand like a newspaper. His death in a car crash some years ago was a great loss to the island.

Up until around 1962 there was no bare boat chartering in the Caribbean. Only a few large charter boats with captain and crew were available. The first bare boat operation I know of was started by Ed Fabian in Charlotte Amalie on St. Thomas in the winter of 1961-62. Until that time the Caribbean had a reputation as a dangerous area for sailing, full of unchartered coral reefs, spawning ground of hurricanes, inhabited by pirates. Suspecting exaggeration, I answered a little ad Ed Fabian ran in *Yachting* magazine. It turned out the boat mentioned had been sold but he had a Cal 24 we could use. A 24-footer seemed a little small for three of us but we made a deal and Harriette, Harry and I flew down by way of San Juan in March of 1962.

Sailing in the U.S. and British Virgin Islands was the best and easiest ever. The weather was always perfect, the wind was steady and reliable, the beauty of the islands and the colors of the crystal clear water were awe inspiring, the natives were genial and helpful. The little 24-footer turned out to be an advantage since it allowed us to anchor in close to shore of tiny islands like George Dog and Fallen Jerusalem. Except for native workboats, we were always the only boat in the harbor. On Jost Van Dyke one man escorted us on a tour of the town and on Salt Island we had long conversations with the local people who were still sending their token rent of one bag of salt to the Queen every year. In Road Town we visited Millie Hammersley who ran the Fort Burt Hotel and had a great lobster dinner with the Bathams on Marina Key. The natives were still building the Tortola sloops, basically by eye without plans; these were seaworthy craft with beautiful lines. On my return I wrote an article for *Yachting* about our trip. This was the first article ever published about bare-boating in the Caribbean and, I'm afraid, helped to trigger the present explosion. Quite a few articles I've written on our adventures to out-of-the-way places have been followed by an

Sailing with the famous Bomba. First bare boat charter in the Caribbean. Ferry in Puerto Rico.

upsurge in tourism. I rationalize my guilt by telling myself that if I hadn't written the article someone else would have. That the ambiance of these places is fragile and must be treated with respect is something I've always emphasized but inevitably some latterday visitor is so supercilious and arrogant that the natives turn sour and antagonistic—and one of the principal delights of these trips is lost for everyone. Fortunately most of the large Carribean charter outfits realize the ecological dangers of the charter boat explosion and try to instill the need for control of littering. In later visits I've been pleasantly surprised at how well almost all charter parties behave as far as dumping garbage and plastic overboard. But, sadly, the local attitude toward visitors has changed throughout the Caribbean and, especially in the U.S. Virgin Islands, hostility is the norm. Examples of thievery and even murder have become more frequent.

The typical boat owner dreams about how nice it would be to have a bigger boat, faster, roomier inside and more comfortable. Olé was great but she was too small to enter such longer ocean races as Newport to Bermuda, Annapolis to Newport and Marblehead to Halifax. In 1961 a new boat was introduced to the market with lines and layout that I greatly admired. The Javelin, 37-foot 10-inch overall, built in fiberglass by DeVries Lensch in Werkspoor in Holland with an Atomic Four gasoline engine, was available in either sloop or yawl rig. Since it had been designed in Bill Tripp's office by Walter Bloemhard, who was now working for me, I asked Walter to keep his eyes open for a Javelin that might be for sale at a good price. In early 1963 he located one in Derektor's yard in Mamaroneck, New York. The boat was called Soufflé and had been purchased originally in 1961 by Talbot Baker who had run into a series of problems: the boat was extremely tender; the overhead was falling off over the captain's bunk; the plumbing in the head wasn't working right; some of the teak trim had come unglued. Acrimonious discussion among Baker, Tripp and Campbell the dealer, Seafarer the distributor and DeVries Lensch, the builders, resulted in Baker, who had sailed the boat for only two weeks, in instructing his lawyers to sell it to the first buyer who came along.

Soufflé had been sitting in Derecktor's shed for two years, so covered with dust that the color of the deck was indistinguishable. On the plus side Walter Bloemhard was aware of most of the problems and knew the cure for them. The boat was very well equipped with nine sails including storm canvas, most of them never used, and virtually every item on the list of optional equipment. I had Derektor add 1500 pounds of lead pigs in the bottom of the water tank right behind the keel and perform other fixes, and I had to settle a bill that Tripp & Campbell owed Derektor before they would let the boat go. But the total cost was still ridiculously low for a full equipped, well designed, almost brand-new ocean racer. Olé was sold for much more than had been paid for her (I had put in much new equipment, a new suit of sails, etc.) and I traded Section 5 to Lamont Haggerty, a fine cabinet maker, in return for a series of extra shelves, drawers and partitions for Souffle, which greatly added to her storage capability and convenience.

Walter's design was very advanced for that era. After we got Soufflé shaken down we consistently beat Bermuda 40s and other boats with higher ratings. The acquisition of Soufflé started a period of intensive ocean racing and cruising. We joined the Norwalk Yacht Club, a good home port, only forty-five minutes from our house but far enough out on the Sound for good sailing and swimming. Some years earlier I had also joined the New York Yacht Club so we used the annual NYYC cruise to Maine to get acquainted with the boat. We also raced in the Off Soundings Club series that fall and enjoyed it so much that we joined the Off Soundings which is quite reasonable as it has no clubhouse or facilities. Its sole function was to run a spring and fall race series, with boats handicapped to their own rule. In 1964 I decided to try to enter the Newport to Bermuda race. Fortunately the head of the Cruising Club of America race committee that year was Tom Watson, who had been interested in our voyage to Bermuda in the Triton and had come on board there. I guess he figured that if we could do it in the Triton we could make it in Soufflé with a crew of seven. Our entry was accepted. The CCA rules for safety equipment are very specific and a detailed inspection is made on all boats to make sure they conform. We went overboard on safety measures; I even posted a large diagram showing all the lockers in the boat which were prominently numbered, with a complete list of all gear, showing exactly where each item could be found. Our menu for the trip was typed up and bound in duplicate in case one was lost overboard; each meal for each day had detailed instructions for preparation with the location of each ingredient called out. The actual cooking and cleaning up was rotated among the crew. A rather elaborate brochure was published in Bermuda giving the particulars of every boat and its crew members, according to a questionnaire sent to all entries. For "cook" I had put down "rotates," so in the brochure "Mr. Rotates" was listed as our cook.

In spite of all precautions something unexpected often happens. In this case Soufflé was robbed at her mooring at the Norwalk Yacht Club ten days before the start of the race. By this time we had everything stored on board except for the food. The thieves made a clean sweep, dumping every piece of gear on the bunks—RDF, radio, clock, barometer, tools, spare rigging items, etc.—and tying it all up in blankets. This is where my inventory list first paid off. I knew exactly every item that was missing so I went to Sears Roebuck and Brewers and re-ordered each item, all covered by our zero deductible insurance policy.

My crew in '64 included my son Harry as starboard watch captain, Shellman Brown as port watch captain, Wells Dow, a roommate of Harry's at Dartmouth, Will Graham of New York, Allen Ames of Connecticut and Peter Hetherington of London. I was skipper/navigator. Peter was a last-minute replacement recommended by our friend, Llew Gibbons, and he turned out to be a very knowledgeable racing sailor as well as a great shipmate.

Harriette had charge of provisioning, including a huge roast beef that was put in the ice box at the last minute and which became a Soufflé tradition in subsequent races. We came in 16th in our class of 21 boats and 62nd in the fleet of 142 on corrected time. Considering this was our first attempt with a new boat, we were happy.

Ocean racing has been described as the slowest, wettest, most uncomfortable—and most expensive—way of getting somewhere. The congestion with seven large men in a 35-foot boat—usually someone else has to move before you can put on your sea boots—is indescribable. In stormy weather, almost guaranteed in the Bermuda race, the act of putting on wet weather gear trousers or going to the head becomes a strenuous athletic feat. The motion is so violent that most people (myself included) go through a period of sea sickness in storms. Watches on duty may have to go up on the fore-deck and change head sails four times between 3:00 and 4:00 in the morning with waves breaking over them and, at the end of the watch, get four hours of sleep before having to get up and do it all over again. The only reward is the faint hope of winning a silver-plated cup which, even if realized, goes home with the skipper. No union man would work under such conditions, no matter how high the pay. It's true

Soufflé, 1963. Cruising in Maine. Dorwin finds a lobster. Harriette at the helm. Dorwin, Harry and friend Jan en route to Stratford (Connecticut) to see Comedy of Errors. *Coming into Dark Harbor. Duck mascot on lookout (when hungry, he would go down to the cabin to get fed).*

there are mitigating factors. Usually (not always) there are a few periods of fair weather, gentle winds, starry nights and sunny days. The competition between the two watches, with the insulting remarks about the sad performance of the other watch entered in the log book, is fun; and there's the comfort of the bunk and the deep sleep at the end of one's watch and the food that tastes so good after four hours of heavy work. Finally, there is the wonderful feeling of accomplishment and relaxation which comes after crossing the finish line. We've done our best and it wasn't too bad. Wives and sweethearts have flown down to Bermuda or up to Halifax for a round of parties and we can re-hash the

Racing in the Off Soundings Club series, spring of '64. Dorwin working on a chart.

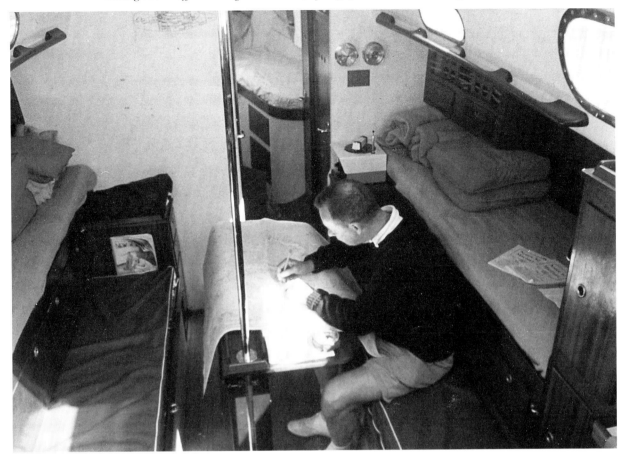

race and compare notes with friends on other boats. If we're lucky enough to win some silver, there are the prize presentations and banquets. Even the skipper can relax; his responsibility is over, all his preparation has paid off, and we are all better friends than we were before the race. This is one of those rare chances for a father to share common experiences with his son; the mutual love and respect Harry and I had for each other became greater than ever before.

One answer to the question—"Why do we do it?"—was answered for me in *The Joy of Stress* by Dr. Peter G. Hanson. Life is normally full of stresses caused by financial worries, family affairs, medical and business problems; these stresses are an important factor in deteriorating health and shortening life expectancy. "The best way to unwind, it has been found," writes Dr. Hanson, "is to switch to something else that is stressful. This alternative stress should be something that requires full concentration, but that involves different circuits of the brain and body. Thus, such obviously stressful activities as roller coaster rides, mountain climbing, white water boating, parachuting, racquet sports, and surfing can all have tremendous value in the reduction of your ordinary stresses." For me, the anti-stress has been in skiing, racing boats and cars, and sailing to little known places.

In 1964 I was still skiing every Saturday but on Sundays I was racing in the "frost-bite" dinghy program at the Norwalk Yacht Club. We raced our Dyer 8-foot prams all winter, with a rescue launch following the fleet to pick up capsized crew. As far as I know, no one has ever been lost or seriously injured in these frost-bite programs which are run by several yacht clubs in Long Island Sound. I also became involved with the Amateur Yacht Research Society. After Walter Bloemhard resigned as secretary for the U.S., I arranged for interested members to meet at the New York Yacht Club to elect a successor. After a lot of fruitless discussion, I volunteered to act as temporary secretary until we could find a permanent candidate. As I might have expected, I was still doing it eight-and-a-half years later, but I met a lot of interesting people in the AYRS, British and American; I enjoyed the free-wheeling attitude of the Society; and actually most of the work was done by my secretary, Mary Smith.

Chapter Fourteen

1963 – 1964

This was flattering since Walt Disney had been making a strong pitch for the job ...

In 1962 the New York authorities decided to host a second New York World's Fair in 1964-65, to be held on the original fair ground in Flushing Meadows, my childhood muskrat trapping territory. The one man who could get things done around New York, Bob Moses headed the Fair effort. A controversial figure, hated by conservationists for his propensity for paving hitherto unspoiled beach and woodland areas, Moses had no equal in being able to force through large-scale projects despite foot-dragging by petty politicians and special interest groups. He virtually invented the modern parkway. Until he proved that it could be done no one had thought it possible that a scenic road could be built, free of hot dog stands, advertising signs and sleazy commercial ventures.

The U.S. natural gas utilities set up a new corporation—Gas, Inc., headed by John Heyke of the Brooklyn Union Gas Company—to put together an exhibit for the Fair. Because of my long association with the gas industry and WDTA's successful record with pavilions and exhibits in the United States and abroad, I was asked to take on the design and organization of the Gas, Inc. effort. This was flattering since Walt Disney had been making a strong pitch for the job. Even after he lost the bid to do the building, Walt graciously agreed to appear in public relations releases with us to help the Gas, Inc. effort. The budget for the building, exhibits, design fees, etc. was set at $6.5 million.

This required the usual period of making organization charts, work schedules and expense allocations, tedious but necessary if the job was to be finished on time and within the budget. Including myself, the key personnel were Bob Blood, Art Clark, Jack Field, Russ Tonkiss for the symbol and graphics, and Joe Mangan as architect of record. Finally we were able to get into the fun part, the actual design of the building and exhibits. My thought was to work out a scheme like Bel Geddes' Futurama in 1939, in which spectator control would be automatic, where visitors would walk into the entrance and be carried the rest of the way past all the exhibits with no further effort on their part. Accordingly, we designed a wide, ascending moving ramp or sidewalk which took visitors up to a large, raised revolving turntable that carried them past the various displays. One of the most obvious applications of natural gas is cooking so right from the start it was agreed that a good restaurant should be a major part of the pavilion. Gas air conditioning was another application with pleasant associations so I decided to enclose the entire pavilion with glass, set in an outwardly leaning angle, with a dark gravel strip surrounding the base. The glass would thus be invisible in the same manner as the invisible-glass shop windows so popular with the better jewelers along Fifth Avenue in the sixties. For the entrance and exit, we planned to use the so-called air curtains which are a moving column of air to act as a screen between the interior and exterior of the building.

After finishing the design and details for the building, the W. J. Barney Corporation was chosen as the building contractor. All the pilings had been driven when we got word from Gas, Inc. that budget collection had fallen short. The new budget was $4.5 million instead of $6.5 million. Most other pavilions

Page opposite: Dorwin with Walt Disney. Below: With John Heyke, Bob Moses and Jinx Falkenburg.

at the Fair were exceeding their original budgets by at least 25% and here we were faced with a 30% reduction. The obvious solution was to cut out the invisible-glass concept entirely and do away with the glass and air conditioning except for the restaurant section and the very plush gas industry club on the upper floor. As the illustrations show, the great overhead canopy turned out to be even more dramatic than the original scheme and the savings helped us to stay within the revised budget. A lot of ingenuity was required to meet the new limit; we worked closely with Barney Corporation to save wherever possible. For example, finishing the underside of the huge roof was prohibitive using conventional techniques requiring elaborate scaffolding and high labor costs. Instead we devised a system for stretching a special weather-proof coated fabric in large pie-shaped panels which were pre-cut by my sailmakers, Ratsey & Lapthorn. Laced in place, these panels looked beautiful and gave no problems for the duration of the Fair.

My concept was to make the building all gas, with the electricity generated by gas turbines. Airesearch in California was selected to supply the gas turbines and generators. We decided on a 400-cycle system rather than the standard 60-cycle alternating current which greatly increased the overall efficiency of the power generation and lighting. Aircraft practice, where light weight and efficiency are paramount, had suggested this. It was a tribute to Airesearch, and our engineering team of Fred Dubin Associates, that the three gas turbines and alternators (one for a spare) performed faultlessly during both years of the Fair. This was one of the first, if not *the* first, full-size demonstration of the potential advantages of a 400-cycle total energy system for large-scale stationary applications—and was most interesting to visiting industry executives.

Every gas utility in the country had its own favorite candidate for the building's restaurant. Whichever one John Heyke chose, he would disappoint all the other contributors. Finally, in desperation, John said, "Dorwin, you're responsible for the building, you pick the restaurateur." Without hesitation I selected Restaurant Associates which had an unequaled record for opening restaurants which immediately achieved multi-star ratings, such as the Forum of the 12 Caesars and the Four Seasons, my favorites in New York. The wisdom of the choice was apparent straightaway in the professional manner with which the RA staff, headed by Joe Baum, went to work. The best and latest equipment was ordered and RA's experimental chef, Al Stoeckli, was

The Gas, Inc. pavilion, the restaurant, the Steinway, the mobile serving cart and the award-winning caster.

appointed to run our restaurant. After the Fair, Stoeckli, who was truly a genius, opened his own restaurant, Stonehenge, which became the best place to eat in Connecticut. Our people selected the furniture for the restaurant and the VIP Club. Russ Tonkiss designed the logo used for the placemats, menus, matchbooks and stationery which later won the Trademarks/USA Certificate of Excellence. We even had a special carpet woven with the little gas flame symbol in the design.

In order to meet our budget many of the furnishings were supplied free by the manufacturers in return for recognition in the handouts and signage where it became, in effect, each manufacturer's "World's Fair exhibit." For the VIP Club we talked Steinway into having us design a special piano; the glass around the Club and restaurant was supplied by American Saint-Gobain, installed with a special new glass hanging technique developed in Germany. Corning supplied a number of specially built propane-powered mobile serving carts, which included a translucent Pyrex panel with a gas burner underneath to keep food warm and a grille where steaks could be broiled right beside the

table, plus a gas refrigerator for cold items. After designing these carts, I couldn't find a decent looking caster for them, so I talked the Jarvis & Jarvis Company into manufacturing a special one to my design and giving us enough casters for the carts, in return for the design right. This caster later won major design awards.

The Korean Pavilion, the symbol for the AIAA.

There were a number of good restaurants at the Fair but the two best, by universal acclaim, were the Spanish Pavilion and the Gas Pavilion. Bob Moses often used to bring VIP visitors to our restaurant in preference to the official World's Fair dining room.

Several months before the opening of the Fair an acquaintance, Paul Enten, came to me for assistance. The pavilion for his client, the Republic of South Korea designed by a Korean architect, was attractive but the best bid he could get for construction exceeded the budget by a factor of two. I asked Josh Barney, our contractor for the Gas Pavilion, if he would be willing to work with us to redesign the South Korean building, using less expensive material so as to keep it within the budget while allowing us both a modest profit. He agreed. By replacing marble with sprayed-on concrete and a number of other cost-cutting measures we were able to bring the building in for what South Korea could afford. The grateful Koreans arranged a formal ceremony with speeches and small gifts for Josh and myself. A few years later Josh would give my son Harry his first construction experience on one of his projects in Newark.

While still at Bendix I had been persuaded to join the American Rocket Society. In 1963 the ARS was combined with the Institute of Aeronautics and the rather clumsy title "American Institute of Aeronautics and Astronautics" was adopted for the joint society. The new AIAA president, Dr. William H. Pickering, head of the Jet Propulsion Laboratory in California, wanted a new logo to replace the separate symbols of the original societies. A twenty-member committee had been selected to oversee the task. A better way to insure mediocrity would be hard to imagine. Already several graphic designers had battered their heads against the committee wall. Each suggestion had been nitpicked to death and the designers had given up in disgust. By this time I had been elected an Associate Fellow of the Institute; somehow Dr. Pickering remembered that I was in the industrial design field so he called me and asked if I would be interested in having a go at a symbol.

Agreeing on a modest fee, I arranged with Lou Theoharides, a private consultant in graphic design, to come up with something. Lou was an ex-employee of WDTA; I always admired his work and I figured, if anyone could do it, he could. After a couple of preliminary discussions, he called to say he thought he had something good. He brought it over, I took one look and said, "That's it." The final art work was prepared and a meeting was set up for us to make the presentation to Dr. Pickering and the formidable committee, which was composed of aeronautical and astronautical engineers, none of whom had any graphic design experience. The reaction was favorable but immediately individual members started to pull it apart. One man said he thought the vertical arrow should be larger; another thought the outer circular ring should be thinner, more and more members began getting into the act. The meeting was deteriorating into a typical committee morass. After about ten minutes Lou couldn't take it any longer. Bolting to his feet, he said, "Gentlemen, I have been paid to design a symbol because I am supposed to be an expert in graphic design. I have given the design a great deal of thought and it is now exactly as I think it should be. You will either accept the design as it stands or forget it entirely. Thank you." There was a stunned silence, a motion was made to adjourn, and a couple of days later Pickering called to say that our design was accepted as presented. In 1998 it is still used in the journal, on letterheads, presentation pins, etc., exactly as Lou first designed it.

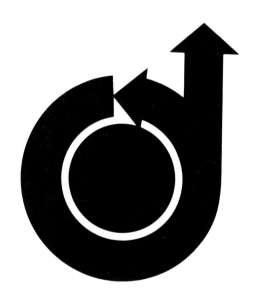

The little carts that we designed for the Gas building restaurant got me interested in the possibilities of infrared gas burners for cooking. A new client, the Columbia Gas System, was one of the most progressive of the large gas utilities. Our contact there was Sy Orlofsky, Vice President in Charge of Research. In the natural gas industry there are no General Electrics or Westinghouses as in the electrical field so the gas utilities themselves do most of the research and development. Our first job for Columbia was a self-cleaning oven. Then I made Sy a proposal for developing a new type of kitchen range—one which would have a single polished flat sheet of white ceramic for the top with infrared burners underneath. Most ceramics are transparent to infrared radiation. The main problem was to find a material which could stand the high temperature. Pyrex was all right for warming-plate temperatures, but, for cooking, the material had to have much higher temperature resistance.

Sy gave us the go-ahead for the smooth-top range idea. We started in typical artist/engineer fashion by thinking first of what the ideal cooking top and controls should look like, long before we had any idea on how to accomplish it. To an engineer this is approaching the problem backwards. Conventional procedure would be to solve the details first and then put them together as well as possible. Our first wood and plastic mock-up was very clean and simple; the controls relate directly to the burners and it is easy to tell which burner is on and by how much.

Columbia Gas liked the concept and said to build a working model incorporating the features of the mock-up. For the most important key to the actual unit, the material for the top, we turned to Corning, our source for the carts. Corning was making missile nose cones which had to both withstand stagnation temperatures of 1500° Fahrenheit and be transparent to radar emissions. A material called Pyroceram had been developed which had a zero coefficient of expansion and could withstand rapid temperature shocks without damage. Moreover, it was quite scratch proof, white in color, and could take a good polish. For the burners we found a small company called American Thermocatalytic, financed by Carl Loeb of the banking family. Its burner was a proprietary, gray, fuzzy-looking material, vacuum deposited on a stainless steel screen through which a gas/air mixture was forced. When lit the burner glowed brightly. About 70% of the heat output was in the form of infrared radiation.

During the early stages of the development of the working model Sy Orlofsky and I were at Corning one day, having lunch in the executive dining room, when the Corning president calmly announced that they had decided to go into production on a unit such as we had designed, but with electric instead of

The smooth top range for Columbia Gas. The low pressure blower (above), the gas burners (below), Dorwin during the testing phase.

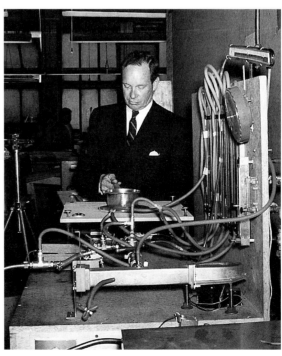

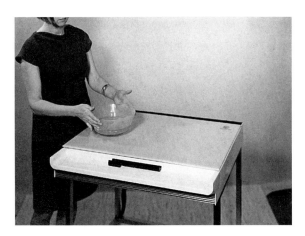

Columbia Gas agreed that publicity for the smooth top range would be useful. So James Beard, the best-known chef in America, was enlisted to cook some dishes on the prototype in the WDTA lab. The staff was amazed at the gourmet treat he made out of ordinary hamburger on the new range.

gas burners. This was a bad blow to Sy and Columbia Gas, who had financed the development for the good of the gas industry. But there was nothing we could do about it. Many thousands of Corning's cook tops were sold at premium prices. The electric burners were very slow compared to natural gas but people were receptive because of the attractive appearance and ease of cleaning.

We quickly found a substitute for Pyroceram called Cervit, made by the Owens-Illinois Glass Company. Cervit could withstand even higher temperatures than Pyroceram and Owens-Illinois was very cooperative in helping us. After developing a special round, concave American Thermocatalytic burner, I designed an electrically-driven blower to supply the necessary air pressure, plus a control system which worked like the push-pull controls of the mock-up. The working prototype looked almost exactly like our original wood and plastic mock-up and performed beautifully. Unfortunately, the gas version never went into production. Twice different companies have been on the verge of producing it. Tappan reached the stage of holding an elaborate press conference at the St. Regis in New York, but something interfered at the last minute.

By this time I had quit automobile racing and traded in my Porsche Speedster for a more comfortable model called the Convertible D. But what I really wanted was a new 365SC Porsche and in '63 it was possible to save more than a thousand dollars by buying one in Germany. So the car was ordered to be delivered at the plant at Zuffenhausen, outside of Stuttgart, a couple of weeks before

Above: The Teagues' Porsche Convertible D and Harriette in Stuttgart with the 356SC which replaced it. Below: Dorwin in the Porsche factory alongside the recently introduced 904.

Christmas. Flying directly to Stuttgart, Harriette and I were picked up at the airport by the Scheufelen limousine, which took us back to Oberlenningen where we spent a couple of days touring the various castles and good eating places with the Scheufelens. Klaus also made arrangements for me to visit the racing section at Daimler-Benz, something few outsiders ever saw. This was just before the Mercedes was taken out of active competition.

When we got to the Porsche factory the car wasn't quite ready. The Porsche people were very apologetic and put us up at the Hotel Graf Zeppelin in Stuttgart. A thorough tour of the plant including the racing section was also part of the apology. I was especially impressed by the recently-introduced 904. I still think it is one of the most beautiful cars ever built. After taking delivery of our car, Harriette and I made a leisurely tour through Germany, Holland and France. Because this was the non-tourist season the roads were almost deserted so, although we were restricted to about 90 mph for the break-in period, we could make good time. And there was no problem getting the best rooms at the inns and good tables in the restaurants. Our vote for favorite place was the Ferme San Simeon at Honfleur, one of the most attractive towns in France. From LeHavre—

where we left the car to be shipped home—we took the express train to Paris to do our Christmas shopping and saw the Marx Brothers' *A Night at the Opera* before catching our plane for home.

Early in 1964 Norman Cousins of the *Saturday Review* called about a design issue he was putting together for May. What he wanted from me was an article delineating what I considered the best industrial designs in the world since World War II. This sounded like fun. Here was a chance to express my appreciation for a number of products I admired via one of the most respected periodicals in America. (The article is reprinted in the Appendices.) I resisted the chance to include one of my own designs but I actually owned several others—the towel rack, Scott ski poles, an Ericophone, some Wegner chairs, and a Triton sailboat. Obviously most of these designs have been superseded but all still look aesthetically attractive. The article received lots of publicity, including a follow-up exhibit at the Ringling Museum in Florida.

That December I got a phone call from Larry Braymer, head of the Questar Company. Although I was a bit busy at the time, my attention quickly revived when I realized he was offering to give me a Questar telescope for Christmas.

Chapter Fifteen

1965 – 1966

At the Operakalleren the head waiter turned out to be a fellow Ellington buff ...

Upon learning that various government departments and agencies were looking for less costly ways of putting on foreign displays, I approached Commerce with an idea: one basic design to be made in modular exhibits that could be dismantled and shipped from country to country. Commerce liked the approach and we designed such a system which was used in many European trade fairs. We did a similar thing for Agriculture. During the summer of '65 I felt it was time I visited some of these exhibits, and Harriette and I decided it would be fun to get in a little small boat sailing along the way.

Our first stop was Berlin for a Department of Agriculture exhibit. After visiting the trade fair and meeting with officials, we rented a car and drove down to Wannsee by the famous Avus track which before World War II was the fastest track in the world, allowing speeds of over 250 miles per hour. Wannsee is on a bay in a lake called Havell. During lunch the head waiter told us of a small boat club that had been organized to rent daysailers to American military officers but was now being phased out. After some searching we found it. For a modest reward the custodian had no scruples against letting us have a boat. It was a warm, sunny day with a nice light breeze so we sailed around the lake for a couple of hours before returning to our hotel in the city. West Berlin was back in business as usual, full of smart shops, the people mostly well dressed and looking prosperous. Harriette crossed Check Point Charley into East Berlin with a tour one day while I was working; she told me that her bus was driven over a large mirror before returning to make sure there were no East Berliners clinging to the chassis underneath.

Dusseldorf was next. We met with George Gardner and others in my department who were setting up an exhibit at the Fair there. At lunch in the baroque Park Hotel, I ordered a dry martini for Harriette until George Gardner reminded me that I had just ordered three (*drei*) martinis instead. We also visited fairs in Essen and Cologne before crossing into Austria.

In Vienna there was a meeting of the International Congress of the Societies of Industrial Designers (ICSID). Because Harriette had never been to Vienna before, I had booked rooms at the Sacher, one of the most fascinating hotels anywhere. The interior is full of little special dining rooms and music rooms on various levels; I found one delightful little wine and coffee room, with a zither player, approached by a winding staircase that I had missed completely on my earlier visit. We had the obligatory *Sacher-torte* and *kafee mit schlagsahne* in the sidewalk cafe and visited the many sights of Vienna. My friend Bill Turner, an ardent horseman, had asked me to buy him one of the famous Austrian saddles but the saddlery I approached was astounded that I wanted to buy a saddle without bringing along the horse and I had to forget that assignment.

About fifty kilometers southeast of Vienna is a lake called the Neusiedlersee, one end of which is in Hungary. We drove down to the town of Rüst on the shore of the lake and, sure enough, there was a marina which rented small daysailers, their mainsails emblazoned with the name of a famous European cold creme to bring in a little extra revenue. Again we were lucky with the weather, and we had a pleasant sail down to the south end of the lake and back where we might even have crossed into Hungary for awhile.

Our next stop was Stockholm to check on the trade fair there. Our room in the Grand Hotel overlooked the royal palace across the lagoon. On a previous visit to Sweden I had sailed in the archipelago off Saltjobaden which is reminiscent of Maine and quite beautiful. Although it was the first of October, the end of Stockholm's short summer season, I wanted Harriette to see the islands if possible. We rented a Volvo, drove to Saltjobaden, checked out several clubs and marinas, and finally approached a teenager whose English was quite good and who was working on his family's sailboat that had just been hauled out. Chartering a boat was going to be difficult, he said, because all the owners he knew were getting their boats ready for the winter. So, regretfully, Harriette and I headed back to our hotel.

Several kilometers down the road a VW Beetle drew alongside, its horn honking furiously. It was the teenager with his mother and father. He had told his parents about us, they had quickly decided we

The reception in Stockholm, the charge d'affaires and economic advisor to the U.S. embassy on the left, Harriette and Dorwin enjoying an hors d'oeuvre on the right. Below: Skiing in the Catskills over the winter of '65–'66.

shouldn't miss a sail in such pleasant weather and could just as well haul their boat out another day. After some half-hearted protests we turned around, went back to their boat which the Nystroms had put back in the water and had a lovely sail to their summer home on one of the off-shore islands. Late in the afternoon we took the Nystroms out to dinner—a pleasant end to a pleasant day. This sort of hospitality could only happen in Scandinavia.

We also visited the resurrected *Wasa*. The ship had sunk in the 17th century and, after 300 years under water, had recently been raised with all sorts of artifacts intact. One night we had dinner at the Operakalleren, Sweden's most elegant restaurant, where I got to talking about jazz with the head waiter who turned out to be a fellow Ellington buff.

Back home that winter we did a lot of skiing at Highmount in the Catskills often taking a narrow country road called Tweed Boulevard which ran along the top of the Palisades near Nyack. Along the way we noticed an interesting contemporary house under construction with a "for sale" sign in front and decided to take a look. Built with four levels on

The house at 155 Tweed Boulevard in Nyack designed by Charlie Winter. Photographs of the house from the road, from the side, from the back. The Teagues moving in, with Volkswagen, Porsche 356SC and station wagon.

a very steep slope, the house had been designed by Charlie Winter, a Nyack architect, and was being built on speculation by a local builder named Ralph Berechid who was getting a bit worried as few people seemed interested in it. The nearest neighbors were a quarter of a mile away in one direction, three-quarters of a mile in the other. The elevation was about 300 feet; the house overlooked the Hudson on one side and park land on the other.

It seemed a shame to move out of Glen Goin with its low rent and other attractions, but it was time we owned our own house and the Tweed Boulevard place was so attractive that we bought it. Chase Manhattan Bank, where I had my account at the time, refused to give us a mortgage because the design was too radical but we found a local bank that went along. The house is now worth several times what we paid for it. The decision to move was

really serendipitous because shortly thereafter Mrs. Rionda died and the whole Rionda estate was left to a young friend of hers who lost no time in selling it to developers. Shortly after we moved to our new home we attended a round-robin party where everyone visited ten or twelve houses in Nyack and Sterling Forest designed by Charlie Winter.

By this time WDTA had a new project in the office that involved a consortium of oil and gas companies organized for the purpose of setting up a community pipelining system in the Gulf of Mexico. Normally each individual company had its own pipeline system from the offshore platform to the shore, resulting in an underwater spider web of pipes. The proposed system was called the Red Snapper Project and envisioned a single 30-inch pipe, 120 miles long, running parallel to the shore with two 30-inch feeder pipes located about 60 miles

apart running into dual refineries where the crude oil would be refined and the gas and oil would be separated for shipment and transmission. Each of the twenty or so participant companies would be connected into the 120-mile line with their contribution to the total metered individually. Both gas and oil was to be fed into the system. The gas would be moved by "sphering"; huge rubber "basketballs" that just fit the pipe interior would be fed into the line at spaced intervals and would sweep the mixed oil and gas along until it reached the refinery. This was a daring, innovative concept that promised to greatly increase the efficiency and reduce the installation and operating costs of all the participants, but it required much sophisticated engineering and ingenuity in order to succeed. Columbia Gas was a minor partner compared to such giants as Shell Oil and Mobil but the Columbia Research Department headed by Sy Orlofsky was carrying the largest part of the R&D load, out of proportion to their share of participation. Columbia engineers had just recalculated the wall thickness of the 30-inch pipe using a more sophisticated mathematical analysis and found that the wall thickness could be reduced by a sixteenth of an inch. This doesn't sound like much but it resulted in an estimated saving of several millions of dollars. This gives some idea of the scope of the project.

Sy Orlofsky brought in my group because of our record of innovative and successful solutions to several prior problems. This one was especially interesting to me because much of the work was involved with underwater operations to depths of over 300 feet. The scope and high stakes associated with the oil and gas industry operations in the Gulf of Mexico are difficult to comprehend. For example, the company that serviced the offshore platforms had a fleet of helicopters which was the third largest fleet in the world after the United States and Russian air forces. As part of our indoctrination we were flown out to the oil and gas rigs in Houston and New Orleans. From the start it was apparent that major improvements would be required in pipe laying, adding or capping off lines, in-place valve installation, etc., in greater depths than previously encountered. We had a series of meetings with Shell Oil's research group; MacDermott, supplier of the so-called "lay-barges" to put down underwater pipe; North American and Grumman to assist with engineering; Andre Galerne of the International Underwater Contractors; and Bill Bascom, head of Ocean Science and Engineering—all acknowledged world experts in their respective fields.

Charlie Fromm of my group handled the liaison. Walter Bloemhard was in his element, proposing all sorts of ingenious platforms, underwater habitats, multi-hull work craft and novel ideas for working at

Nov. 25, 1969   W. D. TEAGUE, JR   3,479,831
METHOD AND SYSTEM FOR LAYING PIPE UNDER WATER
Filed Sept. 20, 1967   3 Sheets-Sheet 1

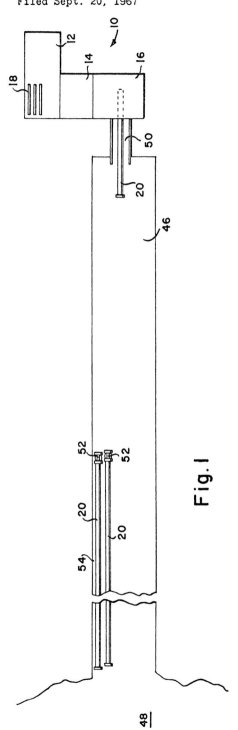
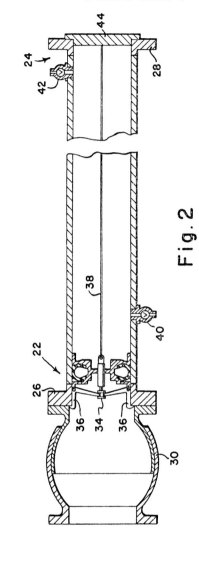

INVENTOR.
Walter Dorwin Teague, Jr.
BY
Curtis, Morris & Safford
ATTORNEYS

The patent for method and system for laying pipe under water, filed by Dorwin during the Red Snapper project. Dorwin was interested in the project because it involved underwater operations to depths of over 300 feet. Above: Soufflé in storage and in the yard. Below: Sailing in '65–'66; Harriette (left); Dorwin flanking Bill Shea and Ben Kochade (a friend of Klaus Scheufelen's) during a sail. Soufflé took third place in the '65 Annapolis/Newport race.

great depths. We met in New York with a representative of the Hawaii-based Oceanic Institute, headed by Tap Pryor (son of Pan Am Senior Vice President Sam Pryor), who owned a vast estate on Maui including the Seven Falls. We were impressed with Oceanic's program for developing an offshore underwater research facility on Oahu which featured an underwater habitat that could be reached via a tunnel from the shore. Accordingly, I set up a meeting in Hawaii with Sy Orlofsky and brought along Sy Gollin from our Anaheim office to handle future liaison with the Oceanic Institute. After a thorough briefing on the program by Tap Pryor we were given a tour of the Oceanic facilities as well as Sea Life Park, a public exhibit featuring trained porpoises and orcas run by Tap's wife (a daughter of author Philip Wylie). After agreeing that the Oceanic Institute would be a worthwhile test facility for the Red Snapper program, a meeting was set up for January of '67 for Tap Pryor to make a presentation to the entire Red Snapper engineering group.

During the summers I was racing Soufflé and in 1965 took a third in our class in the Annapolis/Newport Race. In 1966 the Cal 40 was the hot new boat and I decided that it would probably win overall honors in the Newport/Bermuda Race. Bill Green, a friend of my crew member Tom Nolan, knew of a Cal 40 whose owner wanted very much to enter the Bermuda Race but didn't have the necessary qualifications and was willing to charter it to us for a dollar providing he and his son could come along. I took him up on the offer. We made arrangements for the boat to be brought up to Newport for final preparations. As I suspected, a Cal 40 won overall honors, another Cal 40 was close behind the winner but our boat was farther down in the fleet, due mainly to my error of staying some distance west of the rhumb line, based on the history of previous races. Also the idea of having the owner and his son on board, but not in charge of the boat, turned out to be a mistake. This was one of the few races for me that ended on a rather sour note.

Chapter Sixteen

1967 – 1969

The new firm's location was decided by finding the epicenter of the hometowns of the people involved ...

By 1967 the political situation at WDTA was rapidly becoming impossible. Milt Immermann, the salesman we had hired who was acting as an account executive, had few qualifications as a designer in my view and operated more by ingratiating himself with the client's top executive and giving the client what he thought he wanted, rather than trying to get the company to adopt good designs. Bob Harper, WDTA president, was falling ever more under Immermann's influence. Surprisingly, Jack Brophy, who was normally ultra-conservative, was also being influenced. It was becoming quite common for me to discuss some decision or procedure with Bob Harper (I was still ranking number two partner), arrive at a clear-cut agreement with him and, on my return from Europe or California, find that the decision had been reversed and a completely different course was in effect. To have this situation corrected, I would have to spend a large percentage of my time in company politics instead of on useful projects.

I discussed this with Bob Harper and even warned him that, if I ever left, he would be no match for Immermann's conniving. He dismissed the idea and said he had no worries about handling the man. In addition to my frustration with the way things were going at WDTA, I was becoming increasingly disenchanted with the daily commute to Madison Avenue. This was not only expensive (mid-town daily garage rates were equivalent to what a small apartment cost a few years earlier) but the average time for the trip, morning or evening, was up to an hour and a quarter. I was spending over ten percent of each working day on the Palisades Parkway and the West Side Highway. The alternative was to move to Manhattan but I couldn't conceive giving up the many advantages of our rural location to live in town.

Matters came to a head one day during a meeting that had been called with the partners and with our corporate legal advisor, Milton Gould. (This was the same Milton Gould who would be in 1986 headlines for representing Ariel Sharon in his suit against *Time*.) The preliminaries are vague to me but there was some criticism of an action I had taken and suddenly I knew this was the time to get out. My next words are quite clear to me. I got up and said, "I believe I can save some time here. I can no longer work in the same organization with Immermann, who I consider to be without any design ability or integrity. I hereby resign from WDTA, effective immediately." Gould said, "I am flabbergasted," or words to that effect. The meeting broke up.

The next morning when I arrived at my office there was a memo stating that my resignation had been accepted and that I should pick up my personal effects and leave. That night I talked over the situation with Harriette and we agreed that we should go ahead with the Columbia Gas System meeting in Hawaii and enjoy the trip before we worried about what to do next. Leaving WDTA was a rather drastic step for me. I had just bought a new house, I had family obligations, and I had no idea of what sort of compensation, if any, that I could expect from the company. On the other hand, there was a feeling of relief that I had finally broken away and a sort of excitement about starting a new phase of my life. We were due to leave on the following day so I called my good friend Vern Cowper, gave him a brief rundown on the situation and asked if he could arrange a legal firm to help me with negotiations with WDTA.

Harriette and I flew to Hawaii where Tap Pryor not only gave a thorough briefing on the plans of the Oceanic Institute but flew us on a tour of the islands, winding up with a rock lobster dinner at his estate on Maui. We had caught the lobsters ourselves in the reefs in front of the house. The trip was successful from every aspect and the Red Snapper people were unanimous in recommending the Oceanic Institute be part of the program.

On our return I found the Vern Cowper had located the firm of Olwine, Connelly, Chase, O'Donnell & Weyher to represent me. Wirt Marks III was assigned to handle my problem. I had already done a lot of thinking about my next step and decided to set up a new corporation: Dorwin Teague Inc. In the meantime, nearly all of the people who were working on my projects—including my super-efficient secretary, Mary Smith—said they wanted to come along. As it turned out, Wirt Marks was able to make a very suitable financial arrangement with WDTA in addition to an agreement whereby I was

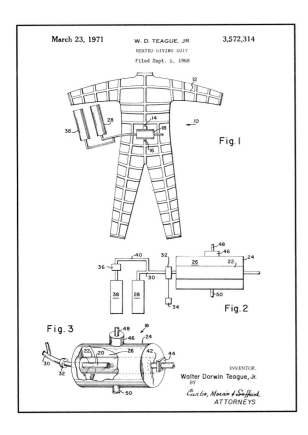

allowed to keep all my accounts. Thus I was in a good position to start off the new organization. The readiness of the people in my group to join the new venture was very touching and there was a great feeling of enthusiasm and optimism among all of us. A location for the new firm was decided upon by finding the epicenter of the hometowns of the twelve or so people involved, which turned out to be Englewood Cliffs in New Jersey. We took most of a floor in a new building on the Palisades which had the added advantage of a marina at the foot of the Cliffs where I could dock Soufflé. My commuting time dropped from an hour and a quarter to an easy fifteen minutes.

I'm sure my leaving WDTA was everything Immermann hoped for. In a very short time Bob Harper was kicked upstairs as chairman, while Immermann became WDTA president and C.E.O. Soon thereafter Harper retired although, fortunately, he kept his position on the board of directors with Jack Brophy and Ruth Teague (my father's second wife). I had very little contact with the company. WDTA went downhill until 1970 when the directors finally came to their senses and forced Immermann to resign. Frank del Guidice took over the presidency. WDTA had come close to bankruptcy but Frank, Sy Gollin and Ed Kibble were able to pull it out and gradually become reasonably solvent. I have never regretted my decision to leave, which gave me a chance to work on projects unhampered by politics and the daily commute to New York City.

Our quarters in Englewood Cliffs were arranged with a unique glass-enclosed conference room and specially designed drafting units. We soon had to take additional space for a shop and a test laboratory to handle new development work. In nice weather many of us ate our lunch in a park with spectacular views of the Hudson and the northern tip of Manhattan across the river.

Unfortunately the Red Snapper project, which looked so promising, was unexpectedly ruled illegal by the Federal Power Commission on the basis that it would give participants an unfair advantage over non-member corporations. This seemed a somewhat specious reason as the concept would have increased the efficiency of the entire industry. But there was no recourse. The project had to be abandoned. I had already obtained two patents on a completely new method for laying pipe in deep water and a heated diver's suit which used bottled gas instead of electricity. Columbia had plenty of other research and development assignments for us and continued to be a major client.

One such project was a new type of natural gas infrared burner. Several ceramic infrared burners were already on the market, mainly for space heating in factories and exterior cold weather applications, which used a molded ceramic tile about a half-inch thick, pieced by a number of closely spaced holes, about 1/16-inch in diameter. The ceramic burner was sealed onto a metal plenum chamber into which a combustible mixture of gas and air was introduced. When the mixture coming out of the holes was ignited it burned close to the surface which became incandescent and produced a high percentage of infrared emission. The ceramic tiles were often surrounded with reflectors to help direct the infrared radiation. Infrared rays are especially effective in large areas or outdoors since they are not affected by air currents and require no circulating fan.

The main problem with these conventional tile burners is that they have a limited turndown ratio. They are either on or off, cannot be modulated down to a simmer level and thus couldn't be used for a kitchen range or other appliance where considerable modulation is required. One assignment was to see if there was some way to make an infrared burner of the ceramic-tile type with a wider turn-down ratio.

This was not the usual industrial design project since appearance was of no consequence. Rather it was an engineering or scientific problem. As a result of my seven years at Bendix, I had achieved some status in the field of air and gas flow dynamics, held a number of patents in the field and had been elected an Associate Fellow of the AIAA for my work on this sort of problem. So I felt that I might contribute something. I will try to describe in layman's terms the thought process that led up to the final solution but, here again, the totally non-technical reader may wish to skip the next few paragraphs.

By the nature of the problem the solution had to be a one-man effort, at least until the basic principle could be established. My first task was to try to analyze why the conventional infrared tile burner was so limited. First of all one must understand that the action of these infrared burners differs basically from that of a conventional gas range burner, in that the mixture of gas and air entering the little holes in the tile has to be almost exactly the correct proportion (about 16 to 1) of air to gas by weight. If too "rich" (too much gas and not enough air), it will not burn on the surface of the burner, which will not incandesce and there will be no infrared radiation. If too lean, it won't ignite at all. The correct proportions are called the "stoichiometric ratio" in scientific parlance. In a conventional gas burner like those in a domestic gas range, the mixture entering the burner is very rich but picks up more air after it leaves the burner and burns well off the surface with the typical blue flame. Nothing wrong with this except that the conventional burner has very little infrared radiation.

The problem with the infrared burners is that the stoichiometric mixture coming into the burner is highly explosive. The explosion is not destructive since there isn't enough mixture to be dangerous but it immediately extinguishes the flame on the outside and inside of the burner. Therefore we have to make sure that no flame or tiny spark ever comes into contact with the mixture in the plenum entering the burner.

About this time some bright student will raise his/her hand and ask, "Sir, what keeps the infrared burner from blowing out the flame? You've got white hot ceramic and flame on the outside surface. Why doesn't it just burn down through the little 1/16-inch holes and set off the explosion?" Ah! Good question—this is the crux of the whole problem.

First you have to realize that a gas explosion like this appears to happen simultaneously, but it really doesn't. If we had one of Doc Edgerton's strobe cameras taking pictures we would see the explosion starting out at the little spark and spreading out at a very fast but very constant rate in all directions. This constant rate is called the "flame front velocity" or "propagation velocity." In our ceramic infrared burner we need some pressure in the plenum to force the gas/air mixture out through the little holes. It doesn't take much pressure to create quite a high velocity in the little holes, in fact it is (must be) higher than the propagation velocity so the flame doesn't have a chance to advance back down against the flow. It's like a man in a four mile-an-hour outboard trying to enter a stream coming out of a lake at five knots. He can never get past the entrance. Now we can see why we can't turn down or modulate these infrared tile burners very far. As soon as we reduce the velocity of the mixture in the holes to the point where it is no longer faster than the propagation velocity—boom! So we are walking a very narrow line—if we get too much velocity in the burner it won't burn on the surface and there won't be any infrared; if we have too little velocity we will get so-called flashback and the burner will blow itself out. Fortunately the amount of pressure required to reach this crucial velocity is not very high but it has to be controlled accurately.

Here we were faced with some apparently immutable physical laws and how could we get around them? How could we get a higher velocity in the little holes but at the same time prevent the flame from lifting off the surface? My next thought was that maybe there was some way to keep the velocity at the outer burning surface low but the velocity in part of the passages high enough to prevent flashback even at reduced flow. About this time the light dawned. Everyone had been thinking of a straight cylindrical hole, 1/16-inch in diameter and 1/2-inch long. If we made the passages in a typical venturi shape rather than cylindrical, we would have our low velocity at the expanded exit and a much higher velocity at the throat. I've always been fascinated by venturis anyway as they provide something for nothing, so to speak. A well-designed venturi can be necked down to one-half the diameter of the straight passage—1/32-inch diameter at the throat with an exit diameter of 3/32-inch and you can get the same flow with the same pressures as the straight hole except that now the velocity at the throat is four times the velocity in the straight hole. Therefore, you can turn it down to one quarter of the original flow rate before you get flashback. And by increasing the exit hole diameter you can handle a greater flow than the straight 1/16-inch hole without "lift off" of the flame.

In the new configuration, the dies cost somewhat more since the many pins have a special shape instead of being straight. But the fact that they are tapered makes the actual molding process easier and less expensive. This is one of the simplest and most basic of the ninety-four U.S. patents I've received to date. It is a far cry from the design of an automobile body, but the approach to the problem and the thrill and satisfaction one gets from arriving at the finished solution are the same. Norbert Wiener, my old freshman math teacher, in his fascinating autobiography *I Am a Mathematician* expresses the same thought with respect to mathematics: "Neither the artist nor the mathematician may be able to tell you what constitutes the difference between a

Dorwin calls this infrared burner patent "one of the simplest and most basic of the ninety-four patents I've received to date."

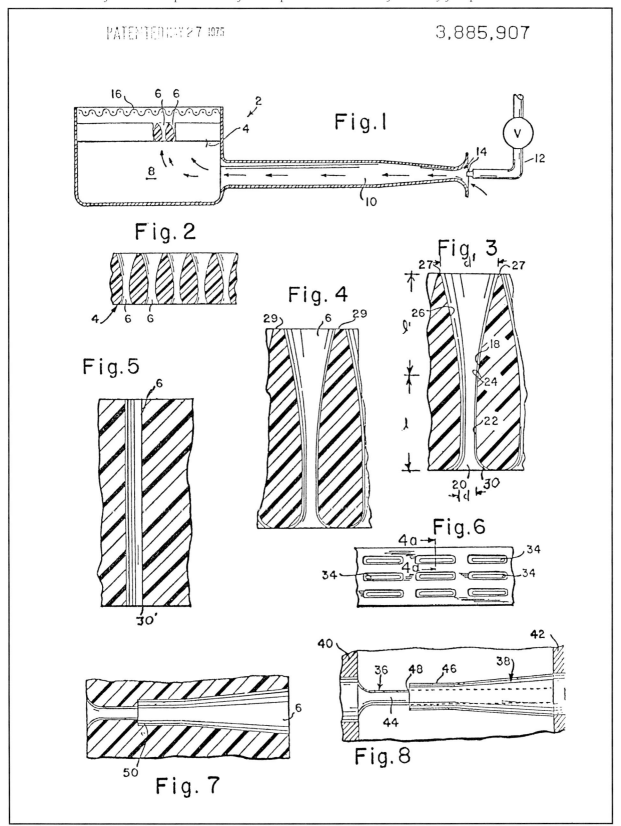

significant piece of work and an inflated trifle; but if he is never able to recognize this in his own heart, he is no artist and no mathematician."

Another project we worked on for Columbia was the Advanced Dryer, to shorten the drying cycle for domestic and commercial gas-fired clothes dryers. Setting up the basic objectives before worrying about a solution, it became evident that two major factors affected the length of time it takes to dry a "typical" load of clothes: lots of heat and the highest possible velocity of the air impinging on the clothes. To these some mechanism must be added which allows all surfaces of each garment to receive the hot air steam which, in conventional dryers, is accomplished by slowly rotating the entire chamber so that the clothes tumble. This was a joint project among myself, Tony Montalbano, Andy Oakes, Keith Osborne and Peter Susey of Columbia. In our final configuration the air enters the drying chamber via three tangential slots and the velocity is high enough to tumble clothes rapidly without requiring the conventional revolving drum mechanism. Our unit was mechanically much simpler but required a large blower and motor. The much higher heat input was no problem with gas; as I remember, we were operating with around 150,000 BTU per hour. Our final working unit could dry a "typical" load of clothes in about ten minutes, compared to forty-five minutes in the conventional dryer.

We had a number of other clients during this period. One was Bill Moog who had been fired from Bendix in 1945 as related in Chapter Eight, and was now head of Moog, Inc., a leading control equipment manufacturer. Another was the Tappan Stove Company which had taken over our smooth-top range; Dick Tappan hired us to design the production version. Tappan also had us work on a rather large project under its vice president for research and development, Dan Meckley. This involved considerable research in time and motion studies and a full-sized mock-up which we built in a temporary rented space adjacent to our office.

A more typical industrial design project was the wall-hung electric broiler for the Nautilus Company in Pennsylvania, which illustrates the effectiveness of the design approach system we had by this time developed to the point where it was used for every new project. In this case our assignment was entirely flexible; about the only directive we were given was to come up with a broiler which would outsell the competition. As I remember, even the choice of power—electricity or gas—was left up to us. Obviously, here was the perfect test for our approach. The client was a small company with limited resources, which couldn't stand a lengthy, expensive research program. We not only had to design an attractive, marketable unit, do 100% of the

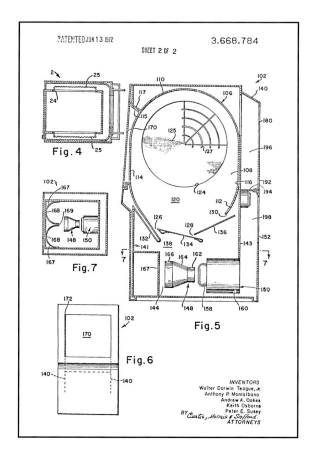

engineering and furnish two working prototypes, but also do practically all of the production engineering which would enable the unit to be built in the client's plant at a competitive price. All of this was handled in our own office and shop, except for sheet metal forming and specially machined parts which we farmed out to local machine shops. The client was very little involved in the development and the final article went into production practically unchanged from our prototypes.

A project as wide open and unrestricted as this

Advanced Dryer; Tappan range, Mary Smith the model.

one would inevitably have gone into a long period of groping through all sorts of solutions without the help of some formal method of approach. Tony Montalbano, Andy Oakes and I headed up the team and Maurice Vauclair was in charge of drafting and production design. As a tribute to our formal approach system, we were able to come up, on the first try, with a unit which was not only successful on the market but which won the prestigious Iron & Steel Award for the best design and best consumer product in all categories for 1969.

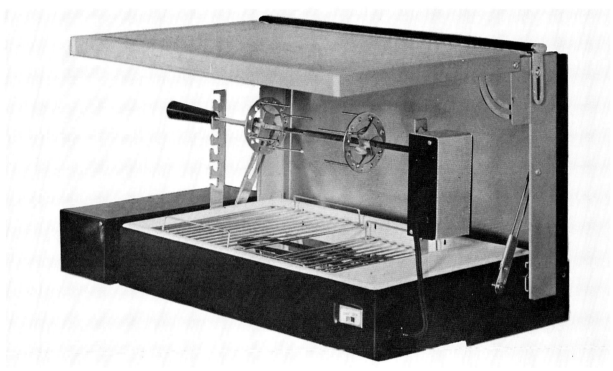

The prize-winning broiler for Nautilus.
The hydraulic servo-valve for Moog, Inc.

A New Yorker *advertisement from 1966; Bachrach was the Teague family's favorite portrait photographer.*
A Volkswagen ad, published after Dorwin declared the Karmann Ghia one of the world's most beautiful products.

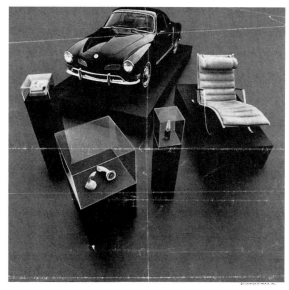

Perhaps the most important feature of the "Teague Approach" (see Appendix A for complete details) was the careful compilation of the list of ideal objectives which came before any actual design work. The client participated in finalizing this list after we had refined our own. This avoided all possibility of misunderstandings later on. So successful was this approach that we were hired to indoctrinate the development engineers of Honeywell in our system and later lectured on it for the Province of Manitoba in Canada.

In spite of all the extra work entailed in starting a new company and setting up a new office—and maintaining a high level of productive effort—1967 was also our most active and successful year in ocean racing. By this time we had Soufflé tuned up to her top potential. My regular seven-man crew, with son Harry and Wells Dow as watch captains, was capable of sail handling and seamanship that compared with America's Cup crew work. Campaigning in New England we won top overall honors in the prestigious Charles Francis Adams Memorial Trophy Race. Next was the three-and-a-half-day Marblehead to Halifax Race in which we were first on elapsed time and first corrected time in Class E. Most of this race was in dense fog which required some tricky navigation past the Bay of Fundy and its immense tides and currents. Following this was the Monhegan Race from Falmouth Foreside, around Monhegan Island and return for our worst result of the year, a second in class, and finally the Chandler Hovey Race in which we won our class and were second overall. This was good enough to give Soufflé runner-up honors behind Ted Hood for the New England Racing Championship for 1967. Skiing and frost-bite dinghy racing in the winter, spring and fall Off Soundings races and cruising the Hudson in the autumn made for an active life but I can't remember being rushed or under great stress at any time.

Nineteen sixty-eight was more of the same, starting with a winter Caribbean bare boat charter. We decided to go back to Soufflé for the Bermuda Race where a minimum rating of 26.5 had been set for calculating corrected time. Soufflé only rated about 25.0. In order to be more competitive, I removed the propeller and faired in the rudder aperture which raised my rating to over 26.0. In the East River on the way up to Newport I jumped up on the bow pulpit to avoid a big power boat wake and crashed down on one of the big bow cleats which badly dislocated my right foot. Steve Swett was on board and brought Soufflé into City Island where Harriette called our neighbor, Dave Andrews, one of the top orthopedic surgeons at Columbia Presbyterian. After a taxi to the hospital Dave and his support team worked for about an hour until the

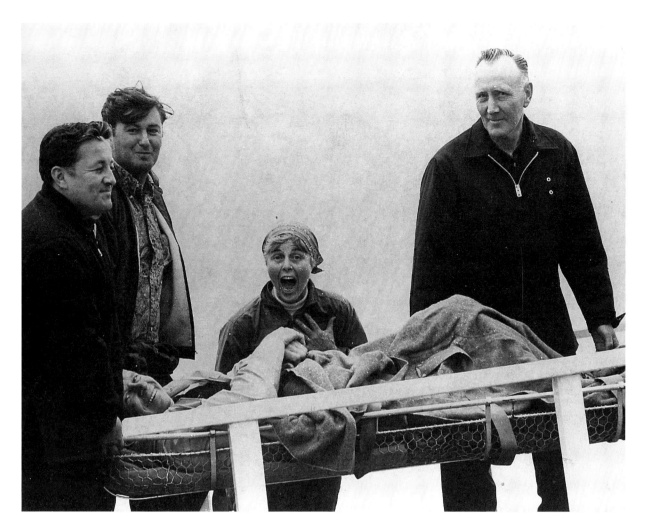

Above: the agony of de-feet? Doesn't seem like it. Harriette appears exuberant and Dorwin is actually smiling as sea rescue workers carry him away after he jumped on Soufflé's bow pulpit to avoid a power boat wake and crashed unceremoniously on one of the bow cleats, which was not the best thing that could happen to one's foot. But Dorwin, fortuitously, is a quick heal. Left: the staff of Dorwin Teague, Inc. One of the perks of employment was the chance for an afternoon at sea. Or river, in this case—an Office Sail on the Hudson in the fall of 1967.

Christmas greetings from Teague friends. Above left: Craig and Barbara Smyth with their children in Italy. The Smyths managed I-tatti, a Harvard-owned museum outside Genoa, during this period. The Smyths boarded Harry during his college-era visit to the country, and were hosts to Dorwin and Harriette as well. Above right: In Vermont, Steve and Sheila Swett and family. Steve was a professional writer and a member of Dorwin's Bermuda race crew among other events. Left: The Teagues' first 911 Porsche, photographed at Martha's Vineyard, 1967.

x-rays showed everything back in the right place, at which point they all cheered.

This was only about three weeks before the race so I had Dave put on a light plaster cast. I sat on top of the workbench in our shop while the cast was fiberglassed to make it waterproof, light and strong. In Newport we had a berth at the far end of the narrow Port O'Call basin; sailing in and out, screaming at the fishing boats that we had no reverse, was quite stimulating. In the race itself, when it got too rough. I didn't trust myself on deck so Tommy Nolan would take the sights with the sextant and I would work them out below. The title I was given by my disrespectful crew was "Bumblefoot" but we managed to place better than ever in the race.

That winter I happened to read an article on British Honduras in *National Geographic*. An accompanying map showed a long barrier reef with a multitude of islands inside. This had to be fantastic cruising territory, and I decided to get more information. No British Honduras embassy or consulate existed in the U.S. and no one I could find had ever been there, except for a friend of Hank Gardiner's at the American Museum who had visited twenty years earlier. Finally, I decided to go down alone and look around. TACA airlines flew there twice a week from Miami and made a reservation for me at the main hotel in Belize, a very primitive little city that gets wiped out every few years by a hurricane. I called on the Governor (British Honduras was then a Crown colony) and met a lot of the local people who told me about a new hotel in the southern part of the country.

Chartering a local Cessna, I flew down only to find that the rumored hotel hadn't been built yet, but I was invited to lunch by the Canadian construction crew. While snorkeling I met a retired

Receiving first in class honors following Marblehead-Halifax in 1967. Below: Sailing off Pulpit Harbor in Maine with the Pools (Larry & Néné), brother Lewis and nephew John. This was Lewis' first sail since contracting polio. Larry, a top surgeon at Columbia Presbyterian, made the trip easy for him.

Eastern Airlines captain who owned a large forty-seven foot trimaran and had been looking for someone to cruise the coast with. It didn't take long to conclude arrangements. Harriette and I flew down a month later for a ten-day trip up the coast, poking into jungle rivers, stopping at many uninhabited islands, meeting Indian families in dug-out canoes and diving along the barrier reef, altogether one of the most fascinating cruises I've ever taken. A lot of the pictures I took appeared with my article on the trip in the March 1970 issue of Yachting.

That summer we cruised on Soufflé in Maine and in October, before we hauled out for the winter, we cruised up the Hudson, very beautiful when the leaves are turning, and visited a number of interesting little river towns like Kingston, Eddyville, Saugertees, Catskill, Athens and Hudson.

Chapter Seventeen

1970 – 1974

At the time the popular concept of conservation was one of discomfort and sacrifice ...

That winter, looking for a new place to sail, Harriette and I decided to team up with our friends John and Amy Stillman on a bare boat charter out of the French island of Guadeloupe. Mail to and from the island was erratic, sometimes three weeks or more in transit. I had responded to a tiny ad in the *Times* by a M. Robert Gautier and, after a considerable wait, finally resorted to the telephone. The boat mentioned in the ad had been wrecked and M. Gautier apologetically offered a slightly larger but much older wooden boat which we accepted. The tiny head compartment had apparently been designed for slim teenagers and some screen door turnbuckles in the rigging opened up under tension but we got it sorted out and stocked up on a vast assortment of fine wines, cheeses and other French delicacies at the Pointe a Pitre *supre market*. We cruised the leeward coast of Guadeloupe and the Isles des Saintes, which were nearly deserted on Washington's Birthday as the French consider them a summer vacation stop. After the Stillmans left, Harriette and I continued on to the island of Domenica where we fell in with Mme. Francis, wife of the chief of the Carib Indian colony, who runs a primitive restaurant there. As usual I defrayed part of our expenses by writing an article for *Yachting*, with pictures we took on the trip.

An architect and well-known ice-boating enthusiast; Ray Ruge owned several ice boats including one of the large antique stern steerers that used to be sailed on the Hudson by the Roosevelts and other old river families. That winter there was some good ice on the Hudson just south of Croton Point, which happens very rarely nowadays with our milder winters and ice breakers keeping the channel open. The Stillmans had introduced us to Ray. One Sunday he let me take out a couple of his boats including the big old stern steerer. Ice boating is a different experience for a sailor, not only because of the speed—50 mph is not unusual—but also because the sails are close hauled on all points of sailing, as the boat speed is so much faster than the wind speed. I could have become hooked on ice boating very easily, but it is an uncertain sport, conditions are rarely right in this part of the world, and another activity was something I really didn't need.

Soufflé was entered in the 1970 Bermuda race. Our blue ribbon crew now included Tip Sempliner, a new employee at DTI. The race start was delayed a day or so because of a tropical storm. We spent the time in practice and a hard-fought croquet match against Carina's crew on Steve Swett's lawn in Jamestown. A day after the race started, we ran into the tail end of a storm with a few indicated gusts up to 60 knots on our anemometer. It was still blowing fairly hard as we entered the Gulf Stream, quite rough as usual. We were wearing a reefed main and working jib. I happened to be at the helm when there was a sharp crack, almost like a pistol shot,

Page opposite: Dorwin and Harriette on a bare-boat charter out of Antigua to the Isle Forché with Steve and Sheila Swett. Above: Soufflé's crew for the 1970 Bermuda race, from the left, Tom Nolan, Wells Dow, Harry Teague, Tip Sempliner, Allen Ames, Steve Swett, the "Skipper." Left: Harry at the helm. Photo: Steve Swett.

and the rod head stay parted about six feet off the deck. Normally the mast would have gone but the wire luff of the working jib saved it long enough for us to lay off and slack the back stay while we took stock of the situation. A spare wire backstay was on board so we kept the full main up to help steady the boat while son Harry went up to the truck, tied in the bosun's chair. He was able to drop the broken headstay overboard and pin the spare backstay in its place, a hairy operation in the rough water, but it went all right. We secured the new headstay with cable clamps that I always carry on board. Within an hour and a half we were back in the race.

For the 1970 race an abandoned Texas tower about seventy miles southwest of the island was made a mark of the course, with the finish on the south shore. This was a safety measure to avoid treacherous reefs north of Bermuda but there was a mix-up about a radio beacon on the tower, and several boats, including a couple of maxis, missed it. The evening before we approached the tower, with the usual Teague luck, I was able to get a three-star fix (one of the few good ones I've ever made) and we sighted it within a few minutes of my predicted time. Even after losing an hour and a half while being pushed in the wrong direction by the Gulf Stream, we crossed the line fifth in our class. Except for the

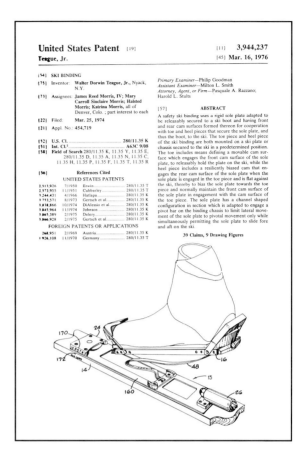

1971 Annapolis/Newport which turned out to be a drifting match where we fared badly, this was our last major race. Our activities thereafter were confined to cruising Soufflé and chartering in southern waters.

Because Scotty was now branching out into other ski products including boots, skis and gloves as well as poles, which were selling as well as ever, I started visiting Sun Valley more often. Russ Tonkiss of our office developed the famous Scott logo, the double "S" that looks like a pair of slalom tracks, among other graphic and corporate identity items. Scotty's was the first of the lightweight plastic boots, patented originally by Franet in California. I designed an adjustable lightweight toggle for the boot that was used for many years. The DTI office also worked on a ski binding invented by Burt Weinstein which, when released in a bad fall, would automatically be reeled back into skiing position by spring-loaded retractable cables. This was primarily an appearance design problem, to reduce the rather bulky effect of the mechanism. A number were sold but it was eventually superseded by the present step-in bindings that are easier to get in and out of.

We were also working with Bill Moog on various projects, including the interior of a disco which Bill decided to build and operate in Steamboat Springs,

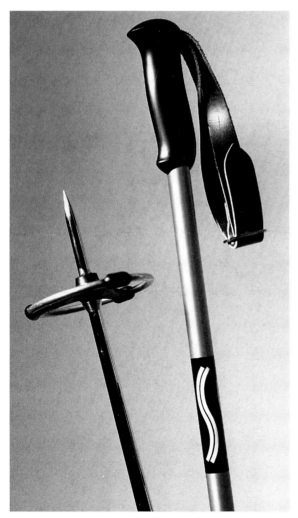

Page opposite: Finish of Bermuda race, 1970. Above: the Scott "S" logo. Ski boot patent.

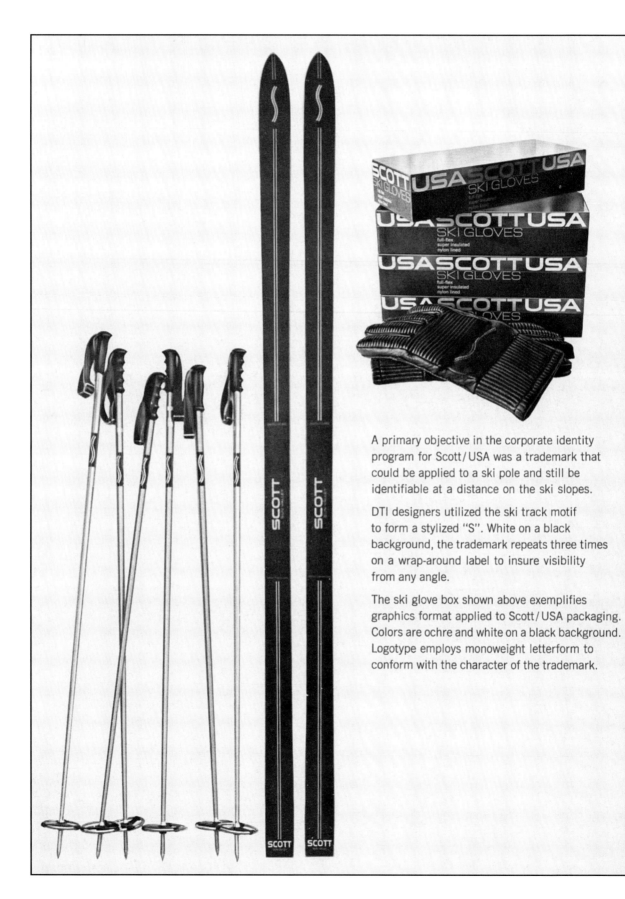

A primary objective in the corporate identity program for Scott/USA was a trademark that could be applied to a ski pole and still be identifiable at a distance on the ski slopes.

DTI designers utilized the ski track motif to form a stylized "S". White on a black background, the trademark repeats three times on a wrap-around label to insure visibility from any angle.

The ski glove box shown above exemplifies graphics format applied to Scott/USA packaging. Colors are ochre and white on a black background. Logotype employs monoweight letterform to conform with the character of the trademark.

Colorado. He had become an enthusiastic and competent skier so he decided to get into the ski binding business. Bill himself came up with a very clever plate-type binding concept that was simple and rugged. The foreign step-in competition finally drove Bill's binding and other plate types off the market although in my estimation the step-ins are much less effective in preventing leg injuries. Unfortunately, the average skier has no way of evaluating the true effectiveness of his/her bindings and depends mainly on advertising claims which may or may not be valid. I developed and patented my own version of a plate-type binding which was partly funded by the Jim Morris family in Denver; the late Gordon Lipe was also involved. This remains the most effective principle yet developed, I believe. We worked out a license agreement with Kurt von Besser in Chicago but he finally backed out due to the aggressive competition of the French step-in bindings and the generally poor state of the ski industry in the U.S.

Our Englewood Cliffs office was not too far from the main offices of Volkswagen of America which also housed the American Porsche employees. We did some graphic design work for VW and I wrote an occasional article for the VW magazine *Small World*. By this time I had bought my sixth Porsche, a 911S Targa, very fast and agile, the last of the pre-emission controlled cars. Porsche had just introduced its 914 model, a moderately priced mid-engine car, but rather ugly compared to Porsche's usually high standard of body design. For fun I designed a new body for the 914—sort of a station wagon effect— that had room in back for a lot of luggage or even a couple of small children. Peter Bishop of our office made some beautiful renderings of it. The U.S. Porsche people were very enthusiastic about the design and urged me to send it over to Zuffenhausen. As I expected, the German Porsche group was not amenable to accepting outside design suggestions.

As a result of our Gas Building for the New York World's Fair, I had become acquainted with Sam Cunningham who ran the Research Development of the Southern California Gas Company (SOCAL). Our work for Columbia Gas was tapering off as it brought more and more of its R&D in-house. SOCAL began to pick up the slack and I was still working on projects for the company until Sam's retirement in 1990. Our first big job was to develop the best answer to the problem of heating multiple dwellings in SOCAL's serving area, where the winters are generally mild. Many apartment owners were using dangerous unvented gas heaters (which are being legislated out of existence). The read-out on our well-tried analysis system indicated a compact, through-the-wall 20,000 BTUH unit, completely sealed combustion with the only protrusion into the

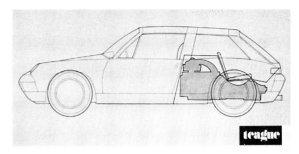

Page opposite: DTI promotional piece. Above and below: The Teague-designed Porsche 914, renderings by Peter Bishop.

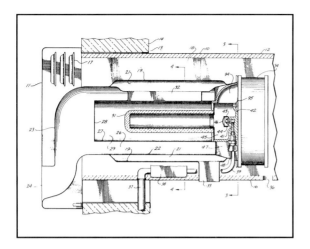

Left: 20,000 BTU infra-red MINICAL heater for SOCAL. Above and below: The smaller version as installed on Soufflé. A Lewis Teague painting is to the right. Dorwin's brother died in April of 1978.

living space being a relatively flat grille. I turned to Cal Loeb's company for an infra-red burner which provided such good heat transfer that the combustion chamber was only a bit larger than a standard beer can. A lot of the development and test was done by new employee and fellow sailor Tip Sempliner, who was very good at building prototypes in addition to having a fine design sense—another artist/engineer type. As the final design developed, Maurice Vauclair did his usual masterful job of producing the drawings. One of Maurice's most valuable assets was his insistence on getting a specific and definite answer on every detail. Prototypes of the heater—dubbed the MINICAL, passed the AGA and life tests, but unfortunately it has never gone into production in spite of strenuous marketing efforts by SOCAL.

    A device that we developed on our own was a simple, propane-fueled infrared cabin heater for boats, recreational vehicles, cabins and spare rooms. A smaller version of the same burner in the MINICAL was used, surrounded by a transparent Pyrex lamp chimney with concentric combustion air inlet and exhaust tubes which act as a heat exchanger to increase efficiency. The combustion system was completely sealed off from the interior of the heated space so the burner could be used in confined areas with windows closed in perfect safety; since its main output was radiant heat the unit needed no circulating blower. The original prototype was installed in my boat where it served us well in cold weather for many years. I received a patent on the concept and took the second prototype to Chicago to demonstrate to Sears, Roebuck. Sears was very interested and, after a long series of negotiations, signed a licensing agreement with DTI for a fair amount of front money plus a schedule for future royalties, an unusual arrangement for the company which usually takes on only well-tried products. Sears had three of its suppliers come up with production designs and cost estimates but had a couple of bad sales years and went no further.

Another new staff member at DTI was Louis Pierre de Monge who had been introduced by Maurice Vauclair. Quite elderly (I never did find out his exact age), Louis Pierre was very alert and full of ideas. His background was mostly in aircraft and automobiles, and he had a long string of basic inventions to his credit. In 1938 Ettore Bugatti had hired him to design and build an airplane to take the world speed record for France. Due to the onset of World War II the plane was never flown. After the war it found its way to the U.S. and wound up in Don Lefferts' restoration shop in Ridgefield, Connecticut, where I had a chance to visit during the preliminary stages of restoration. The Bugatti plane is one of the smoothest, sleekest looking aircraft I've ever seen. It has two concentric, contra-rotating propellers driven by a pair of Bugatti engines, one in front and one behind the pilot, who would lie nearly flat so that the canopy was entirely flush with the fuselage. The plane had been built of laminated mahogany veneers in a Paris furniture factory and a few of the parts were missing so Louis Pierre was acting as consultant to Don Lefferts to make sure that the reconstructed pieces were authentic.

Louis Pierre was very secretive and refused to have his picture taken or to supply information on his past. Only after he had been working for us for some months did I found out accidently that he was the Vicompte de Franeau. He died in 1984—another example of the artist/engineer like his patron, Ettore Bugatti. His beautiful plane is now in an aircraft museum in Oshkosh, Wisconsin—and Louis Pierre is finally getting the recognition he never received during his lifetime.

The 1938 Louis Pierre de Monge racer for Bugatti. It was hidden in the French countryside during World War II. Brought to this country after the war, the plane now resides at the EAA Air Adventure Museum in Oshkosh, Wisconsin. Photo: Jim Koepnick, courtesy of the museum.

Our son Harry was by this time attending Yale Architectural School where he made use of part of his spare time by coaching the Yale ski team. For his thesis he and classmate Peter Stoner designed a hypothetical school for Aspen, Colorado where they had worked during summer school projects. A well-to-do Aspenite, George Stranahan, liked the concept so much that he offered to donate the land if enough support from the town could be arranged to actually build the school. This was finally worked out and Harry and Peter were given the building contract. The construction was completed one summer on a shoe-string basis using volunteer labor and a number of Harry's and Peter's friends from all over the country, living in tents and lean-tos near the site. The project was the subject of a flattering article in *Architecture Plus* which helped launch the careers of the two young architects. The Aspen Community School has flourished and Harry is today one of Colorado's leading architects with a number of impressive residences, hotels and stores to his credit.

It is interesting how the creative genes reappear in successive generations. Harry's half-brother Lewis, while studying moviemaking at New York University, produced an amateur shoe-string film with some friends. A modern allegory on Christ, the title was *It's About This Carpenter* and was good enough to be taken up by public television and shown on New York's Channel 13 several times. This brought Lewis a year's scholarship at New York University but, before he could take advantage of it, Alfred Hitchcock happened to see the film and liked it so much that he offered a job with Hitchcock Productions in California to Lewis, which he accepted. Since then Lewis has progressed steadily. A low-budget monster spoof called *Alligator* which he directed has become a sort of cult favorite among horror movie aficionados. This was followed by a series of successes including *Fighting Back*, *Cujo*, *Cat's Eye* by Stephen King and *Jewel of the Nile*, one of the biggest moneymakers of 1986. His work takes him all over the world, but we all get together for an annual Christmas reunion in Aspen or New Hampshire.

My oldest son Walter Dorwin Teague III is still called "T" by his family. Through our reunions we remain close friends. T is a clinical social worker with a private practice in the Washington D.C. area. He is also an activist and writer who contributes articles espousing otherwise unpopular causes such as our current relations with Cuba. More power to him!

Page opposite: Harry, the Aspen Community School, 1970; a Pennsylvania house, 1989. Above: Lewis Teague (right) in Morocco on the set of Jewel of the Nile *with Michael Douglas and Danny DeVito. © 1985 Twentieth Century Fox Film Corporation. All rights reserved. Below: Walter Dorwin Teague III (fourth from left) with fellow members of the five-member delegation invited to Viet Nam in 1986 by the Vietnamese-American Friendship Association, flanked by Madam Tran Thi Thuong, director of the Ho Chi Minh (City) Food Distribution Co.*

HiJac, the boot remover. Below and page opposite: A "lesson" Dorwin never learned, how not to be intrigued by cars. He was photographed alongside a Marmon Sixteen convertible sedan belonging to Ed Jurist, legendary proprietor of the Vintage Car Store in Nyack, in 1970. The date he sketched on the stained restaurant placemats is not known.

Encouraged by the success of the cabin heater that I had sold to Sears, I began to develop other products in the office, one of which was a new type of boot remover. Gary Van Duersen worked on this with me and I finally got a basic patent on the concept. Ladies' boots were very popular then and I decided to go into production myself on the device. I ordered large quantities of materials and packaging and took some modest advertising space in various ladies' magazines, all of which required a substantial investment of my own funds plus bank loans. Unfortunately the item never caught on, mainly, I believe, because a new, relatively unknown product required much more publicity, advertising and time than I could afford to commit. A couple of years would be required for me to clear up my rather large debts, so I closed the Englewood Cliffs office and moved DTI headquarters to my home with a reduced staff. I hope I learned a lesson from this fiasco. In any case, I have since been able to complete substantial projects at a low overhead, with the help of outside specialists, mostly ex-employees, as needed.

One such project was the Minimum Energy Dwelling, born during the height of the energy crunch. At the time the popular concept of conservation was one of discomfort and sacrifice. For many years architects and builders had paid relatively little attention to energy conservation in housing, courses on energy conservation in architectural schools were superficial, and the popular philosophy was that fossil fuels were an

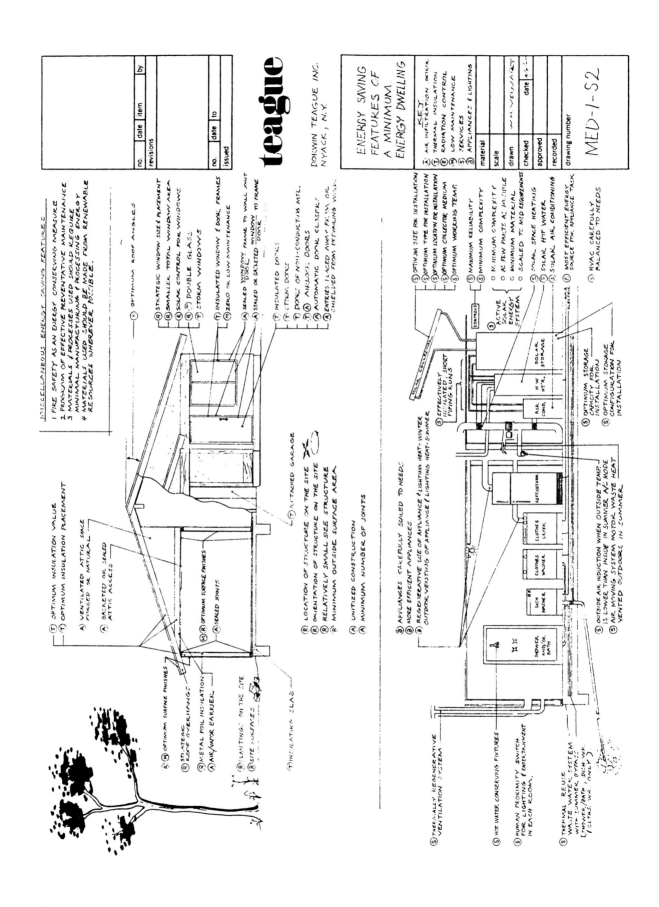

inexpensive and inexhaustible resource. American cars became larger and less fuel efficient every year. Suddenly long lines at pumps made the general public aware that the American standard of living was dependent on fragile international agreements, half of our petroleum was being imported, most of it from notoriously unstable political regions, and even our own oil and gas would be exhausted before the year 2000 if we kept on accelerating our wasteful habits.

Since the early fifties our family had been driving small cars—VWs and Porsches—not for reasons of economy but preference because they were faster, more controllable, easier to park, and more fun to drive. As for housing, I was sure that a fuel-efficient home need not be less comfortable. On the contrary, with enough thought and ingenuity, we could design a very efficient dwelling that would be more attractive and more comfortable to live in. The best way to bring this to the public's attention would be for one of the utilities to build such a house, in which a number of hitherto neglected factors would be taken into account, as a demonstration to show what was possible. I prepared a preliminary prospectus for such a project in which I suggested that some of the funding might be obtained from the Federal government.

My approach to Columbia Gas went nowhere because the company mistakenly believed the project was mainly concerned with active solar energy systems which, it was felt, were being adequately developed already. I then talked with Sam Cunningham at the Southern California Gas Company who became very interested. The first thought was to get several other utilities to collaborate in the project but this didn't work out so SOCAL finally decided to go it alone. In the meantime, I had discussed the project with friends at Honeywell who tentatively agreed to furnish the necessary instrumentation at cost as part of their own research. A new Federal agency—the Energy Research & Development Administration (ERDA)—had been established to foster acceptance of new energy-saving technologies. I was told that the Minimum Energy Dwelling, or MED as it became known, might qualify for funding.

The main stumbling block, which took well over a year to resolve, was to find a builder or developer who would be interested in collaborating with SOCAL. Not until we finally happened to get in touch with the Mission Viejo people did the project really get underway. A wholly-owned subsidiary of Philip Morris, Mission Viejo is a large residential planned community south of Los Angeles which in 1975 had 32,000 residents on some 11,000 acres of property. Unlike most developers, Philip Morris was extremely forward thinking and had an open mind for new techniques which could improve its houses and enhance its already excellent image with the general public. Philip Morris homes were so much in demand that when a new tract was to be opened, prospective buyers would camp nearby for days to be early in line to buy.

The final project was for two MEDs, one a demonstration home open for inspection by architects, builders and prospective homeowners with the many energy-saving features highlighted, the other a nearly identical house lived in by a "typical" family, thoroughly controlled and instrumented by Honeywell so that the actual energy savings could be recorded over a year's time. The layout was based on one of Mission Viejo's most popular units, an 1100-square-foot house called the "Cordova," so that the energy savings could be compared directly. An active solar collector system was included, but this was actually secondary to the sixty or more passive features, a number of which were new with the MED. The total cost of the project including the fairly extensive public relations and advertising campaigns was about $620,000, of which ERDA paid 37%, SOCAL 40% and Mission Viejo 23%.

Collaborating with me on the MED was Wes Verkaart, a new employee, a fellow automobile enthusiast specializing in Saabs, another artist engineer type, very clever in design and an excellent workman and model builder. After leaving DTI, Wes went with a pharmaceutical and medical supply manufacturer where he developed and patented several very innovative and profitable pieces of equipment. At this writing, he and another chap have started their own company, designing, manufacturing and marketing high tech medical equipment. Wes is one of several DTI alumni who have successfully gone into the development of new products.

The results of the MED project fully justified my original concept. The yearly energy saving in winter and summer was over 50% for the passive system alone and the active solar collector system was scarcely needed. After some time the livability of the houses was considerably enhanced over the standard models of the same house. MED dwellings were noticeably easier to keep clean because of less infiltration; another unexpected benefit was a marked decrease in sound from the outside. A well-publicized campaign drew many architects, builders and interested homeowners through the demonstration house and SOCAL received plaudits and commendations from the State and Federal governments for an important contribution to national energy conservation. Mission Viejo also received a lot of favorable publicity and a number of MED features were adopted in future production housing. Both homes were ultimately sold to private

Chapter Eighteen

1975 – 1986

The variety of my work, if anything, increased as I approached my late seventies ...

By 1975 I had over eighty U.S. patents in my name, many of which had been processed by my New York patent attorneys, Curtis, Morris & Safford. One of their clients, the Dictaphone Corporation, was being sued by North American Philips for infringement of a design patent on a portable recorder and, as a prominent designer, I was asked if I would be interested in acting as an expert witness in the case. This seemed like a challenging change of pace from the design business, besides being quite lucrative, so I accepted. In meetings with Greg Neff and his assistants at CM&S, and Dictaphone corporate counsel Bob Falise, a defense strategy was worked out. Before accepting this assignment, I had reviewed the two designs in suit and satisfied myself that the Philips unit was a rather undistinguished design which followed certain standard design clichés of that era, and that Dictaphone had made a sincere effort to make the appearance of its unit different from Philips'. I worked up a series of large poster-size charts to emphasize the detail and overall variations. One of our important exhibits was a very incisive article on popular design clichés written by one of our employees, Gifford Jackson, for *Industrial Design* magazine some years earlier.

The court's decision was that Dictaphone did not infringe. Moreover, the Philips patent was declared invalid; the company would have been much better off never to have brought the suit. My testimony was cited as an important factor in the decision. There was some publicity on the case and, therefore, at the age of sixty-five I was launched on a new career as an expert witness. Fortunately all of the cases I accepted have been decided by a judge rather than a jury. Design patents are notoriously difficult to define and adjudicate; I shudder to think of trying to instruct twelve lay persons in the intricacies of design patent law. Indeed, design patents are new ground for most judges who, in spite of their legal experience, have a hard enough time assimilating all of the nuances of this field. Unlike attorneys who must do their best to defend their clients regardless of the merits of the case, I always had a chance to review both sides of the suit and to turn down those where I felt the plaintiff's design had been obviously copied. In fact I have turned down several potentially lucrative cases because I couldn't be comfortable trying to argue in court that there was no infringement. In most cases I've represented the defendant but I was also the expert witness for the plaintiff in a case that was settled out of court in our favor.

I enjoy working out the strategy of cases with attorneys who, contrary to my prior belief, work unbelievably long hours. And I have a knack for testifying in court, especially under cross examination where one must be able to think fast and accurately. I wouldn't want to make my sole career as an expert witness (although there are a number of people who do) but, as an occasional change, it has been a rewarding new experience.

One of my major design projects in the mid-seventies was collaboration with a former employee, Tip Sempliner, who now had his own design/development firm in an interesting restored barn at the head of Little Neck Bay on Long Island. Tip's new client, Lehn & Fink, a division of the Sterling Drug Company in Montvale, New Jersey, was interested in developing new appliances based on utilizing the available domestic water pressure as a power source. Knowing of my experience in the hydraulic field, Tip thought that I might be able to suggest a good approach. My conclusion was that the best solution would be a nutating motor. The nutating principle is the simplest possible displacement motor but the most difficult one to describe. Suffice to say that the one basic moving part is a ball centered on a disc, called a Saturn, for its close resemblance to the planet. The Saturn does not rotate; it wobbles or nutates, and drives a rotary shaft which is the power take-off. As far as I know, the nutating principle has never been used as a main power source; its one big application is for water meters where the output drives the numerical discs that measure consumption. Most of the millions of domestic water meters in the U.S. for the last seventy-five years, at least, use the nutating principle.

As a power source the nutating principle is relatively inefficient as it has high internal leakage. But, for appliance where a lot of water is needed anyway, this doesn't matter too much. For kitchen pot scrubbers, shower massagers, car scrubbers, etc., it is ideal: very compact, relatively slow speed that doesn't need gears, a lot of power for its size, ideally configured for inexpensive molding. The energy of the 50 to 120 psi water pressure is used which is normally wasted every time someone opens a faucet.

The original application that Lehn & Fink decided upon was a water-powered toothbrush. This I felt was the wrong choice, at least for the initial development, since great design ingenuity was required to combine the jewelry-sized molded components and controls in a compact, hand-held device. Also, most of the exhaust water was not used, requiring a return exhaust line, and internal leakage becomes more serious the smaller the motor. I was over-ruled. We finally came up with a design that was more compact than any of the electric toothbrushes on the market. The brushing action was combined with a Water Pic type of spritz action on demand and tests run by the dental lab at Columbia Presbyterian Hospital in New York gave the action a high rating. Two prototypes were built which performed as designed. At this point, however, Lehn & Fink had a change of heart about the project and sold all rights to Bristol Myers. A total of eight U.S. patents were issued on various appliances using the nutating motor principle, plus a large number of foreign patents, all of which are owned by Tip and me. The project finally became victim of a corporate shakeup in the Clairol division of Bristol Myers. Although disappointing, Tip and I had been well compensated for our development work over a period of three years and Tip was able to build up a well-instrumented hydraulic lab at his headquarters.

Another rather amusing hydraulic project came about for the Castro Convertible Company, foremost manufacturer and marketer of sofa beds. Gene Staudt, manager of the Castro plant on Long Island, and I had a pleasant, informal relationship which resulted in at least one new mechanism patent. Normally, my work has been conducted pretty much on a gentleman's agreement basis with, at most, a proposal letter defining my rates. This worked fine with Gene.

One day I got a message from Bernie Castro, whom I had never met, asking me to join him for lunch at his club near Castro's main showroom on 14th Street in New York. This was in 1978, at the height of the new water bed craze. Bernie wanted to be first on the market with a convertible water bed. He had been getting good reports on my work, he said, and figured I was the one who could make it a reality. This was such a mind-boggling concept I was almost speechless but recovered sufficiently to point out that the weight of the average water bed was around 1500 pounds and one of its disconcerting

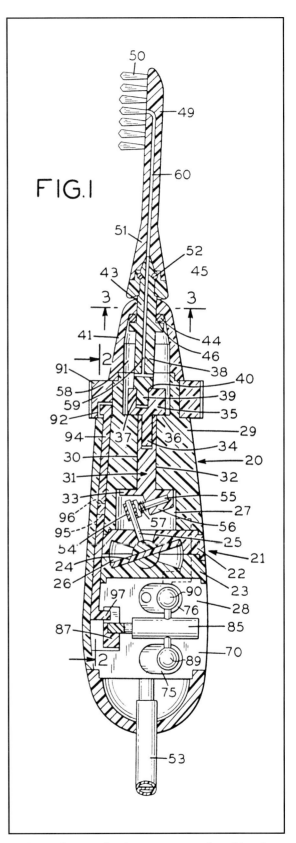

Patent drawing for the water-powered toothbrush.

habits was a tendency to crash through the bedroom floor into the room below. (Water bed dealers sold special insurance to cover this.) Bernie shrugged off this argument as a minor technicality, said he was taking off for his Florida place that afternoon, but would be back in five days and would like me to have some ideas when he returned. This happened to be a relatively slack period, so I spent the next few days canvassing a number of water bed dealers and thinking of possible solutions.

In a week of intensive work, I came up with an idea that actually looked promising. One of the less gaudy Castro convertible sofas was in my office for mechanism studies; like most the line, it was characterized by rather bulky arms and back structure. It dawned on me that the volume of these pieces combined was sufficient for a moderately designed water bed, especially if the ends were made somewhat shallower than the middle. Accordingly, the back and the arms were designed as fiberglass tanks, padded and cushioned as necessary with communicating hoses so that they could be separated for shipping. These tanks held the water when in the "sofa" mode. The "bed" portion was designed as a rubberized tank which, when emptied of water, would be urged by springs to fold automatically into a "Z" section under the "seat" part of the sofa. An ordinary 115 volt submersible sump pump of the type that can be bought at Sears for under $100, coupled to a four-port reversing valve, would pump the water from arms and back into the "mattress" which would slowly deploy automatically as it filled up. Conversely, throwing the lever the other way would cause the pump to suck the water out of the mattress and return it to the tanks while springs in the empty collapsed "mattress" would automatically cause it to fold itself into a "Z" shape under the seat. A suitable proportion of Prestone would inhibit corrosion and prevent turning the bed into a huge block of ice if the owner should forget and leave it in freezing temperature over the weekend. (This actually happened to a friend of mine.)

I showed all this to Bernie on his return. He was quite enthusiastic until I presented my bill for $2,089.00 which I thought was quite a bargain since it included drawings, a patent disclosure and all expenses. At this he blew his stack, swore that he had never authorized any such expenditure and refused to pay the bill. Never had this happened to me before, nor since. Finally, I had to settle for a thousand dollars.

During the early eighties Sam Cunningham of the Southern California Gas Company, which was still actively promoting conservation, asked me to work with their Research Department to devise additional conservation measures. SOCAL's Bill Wong, a capable, pleasant engineer, and I shuttled between Los Angeles and the East Coast on a project which we called by the acronym RECON. Between us we worked up a number of new conservation concepts. One of these was an idea I had in mind from many years earlier which was a controllable reflectance window glass. The windows in a building represent an important source of energy loss, about one-third of the total energy consumption, winter and summer, for a typical Southern California dwelling. A practical way to control this loss would have a tremendous market not only for private dwellings, but for office buildings, public buildings, trains, automobiles—anywhere glass is used and heat control is important. Some rather crude mechanical means have been used to do this, such as the system in the Aspen airport in which the space between the double glazing of the large windows is automatically filled with or emptied of plastic foam beads, depending on whether the sunlight needs to be excluded or admitted to help the heating system. My thought was a coating on the inside of the double glazed window panes which would change its reflectance and opacity by means of a small electric current, in a manner similar to the electro-luminescent effect that displays the numbers in a hand calculator.

With the proper sensors and controls this coating would exclude the sun's radiation in the hot months during the day when it would otherwise greatly increase the air conditioning load, but at night it would open up to encourage re-radiation. On cold days it would admit the radiation in the daytime when it could help or even supersede the main heating system but it would close up automatically to prevent re-radiation to the sky. It should be noted that curtains or venetian blinds are not effective for this purpose since, once the radiation gets inside the window glass, it is too late. Also the various commercial films, although somewhat effective in blocking the radiation, also block it in the winter when it could be saving heating fuel. Because the average homeowner is generally regarded as lazy when it comes to regulating something twice a day, my concept was to have a black box or solid state module which would be connected to inside and outside temperature sensors and automatically adjust the window reflectance or opacity to achieve maximum efficiency winter or summer, in daytime or night. A photovoltaic array in one corner of the window would keep a storage battery charged to provide the small amount of electrical current required. Secondary benefits from the system would be the prevention of fading of furniture and drapes, or the elimination of blinds or drapes if sufficient opacity could be provided to achieve privacy. One can imagine refinements such as a clarity override button so that a parent could check on junior in the front yard.

I prepared a report on this concept complete with a sketch of the physical parts of the system and began to think of the best way to commercialize it. Many years earlier, in the 1950s, WDTA had done a lot of design work on Polaroid cameras and I had formed a high opinion of the company's expertise. It occurred to me that Polaroid might be the ideal outfit to develop and promote my concept. With my usual luck I hit the Polaroid people with the idea at a time when they had just decided to become involved with energy conservation but hadn't yet figured out what product to work on. My idea couldn't have been more fortuitous. They were interested immediately. In a series of meetings it was agreed that a joint venture between Polaroid and SOCAL would be set up, with Polaroid doing the development and SOCAL contributing some of the development funding plus the initial prototype application and demonstration.

SOCAL is a diversified public utility with its principal income derived from sales to around four million customers. Its gas rates are set by the California Public Utilities Commission, so the incentive to make a lot of money on such items as patent royalties is not as great as is the normal corporation. Discussion at SOCAL about the joint venture with Polaroid dragged on so much that the scheduled date for finalizing the agreement was not met. Finally Polaroid decided to go it alone. This was disappointing to those of us who had spent a lot of effort to put the package together. The ramifications of the concept are so vast, when one considers the number of windows in the world that can benefit from this system, and it would have been very gratifying to me to play a part in the development of my original idea.

My friend Scotty had been persuaded to sell out his ski equipment business to a conglomerate and was prohibited from engaging in the ski business. Deciding to stay with sports, he first got into manufacturing back packs, which didn't turn out well, and then into bicycle equipment. At that point he called me in. We spent some time on an infinitely variable transmission and came up with several approaches but finally decided that none would be able to compete with the conventional derailleur. Many inventors have worked on this problem, so far without success. Finally, we decided to go into an improved caliper type brake.

After three years of intensive study and development, the Superbrake went into production and a joint patent was obtained. This brake not only outperforms any other brake, but weighs less than any of the leading competitors. A letter from

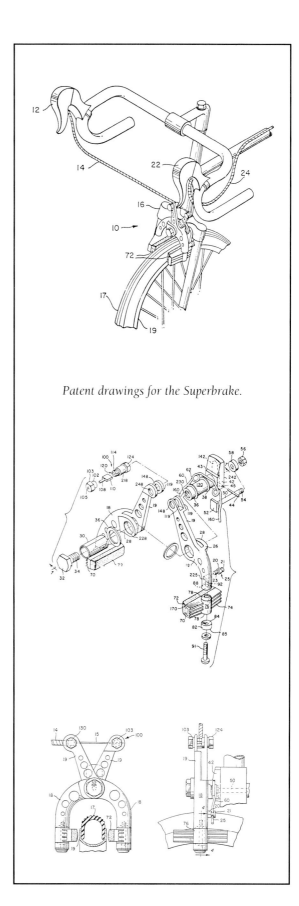

Patent drawings for the Superbrake.

Preliminary layout drawings and promotional flyer for the Superbrake, which Dorwin designed and engineered.

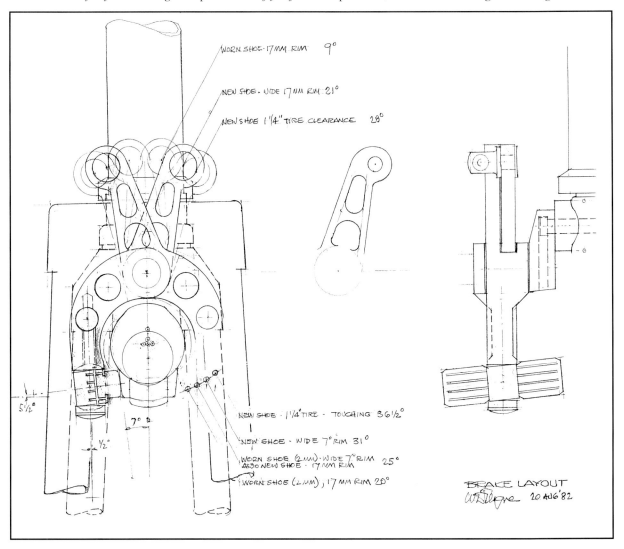

David Gordon Wilson of MIT, a world authority on bicycle and equipment design, provides one unsolicited opinion:

Dear Scotty:

Your Superbrake is magnificent! I've never had such sure and responsive braking! I fitted one brake to the rear wheel of my Avatar, where most (67%) of the weight is, and my stopping ability, and feeling of safety, is transformed. It has a very solid feel to it—no sponginess—no noise or chatter. When I spilled some oil on the rim after oiling my chain a little too liberally there was a quiet squeak until I wiped it off. On my cantilevers, oil produces a loud howl. I don't find either objectionable—the noise could be a safety feature—but the Superbrake is quieter even with the provocation.

As I wrote before, I'm going to type out something about it and offer the other brake with some examples of existing brakes to my fellow design teachers at MIT for them to use as a design case study. I think that it is beautiful as a whole, and the details are equally thought-provoking, delightful and admirable. You are welcome to quote me. MIT would want me to state that I should not be quoted as if I represented MIT, but you may identify me as a professor of mechanical engineering there, and one who teaches design, if you wish.

You deserve success with your Superbrake. You will save accidents and maybe lives. Good luck and thank you!

Sincerely, Dave

The February 1987 issue of *Bicycling*, America's foremost bicycling magazine, included a scientific report on the relative stopping power, in wet and

The New
SCOTT SUPERBRAKE

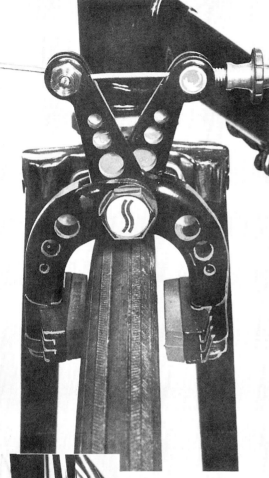

FRONT BRAKE

REAR BRAKE

The same vision, imagination, willingness to discard old outworn designs, and committment to excellence that produced the world's best and most advanced ski pole (in 1960) now have produced an equally-superior caliper brake.

Light Years Ahead of all Existing Caliper Brakes!

For cyclists who want the *newest and the best*.
The nearest anyone has come to an *"ultimate brake"*.

- *STIFFER*
  Much more resistant to flexing and twisting than even the top racing brakes such as Campagnolo. Comparable to cantilevers.
- *LIGHTER*
  About 100 grams per set lighter than Campagnolo and most other side-pulls.
- *NEATER AND MORE COMPACT*
  Fits almost entirely inside the fork and steering tube outline, yet fits 1-1/4" tires (and smaller).
- *SYMMETRICAL*
  Avoids centering problems of lop-sided side-pulls.
- *GREATER REACH RANGE*
  40 to 54 millimeters with one size brake. Extender bushings available to reach up to 60 mm.
- *CRISP, SOLID FEEL*
  Like the brakes on a good sports car. Control by how hard you squeeze, not by how far you compress a soggy lever.
- *MUCH LESS "SPONGE"*
  So little that you no longer need a quick release. Ride with plenty of shoe clearance, so a broken spoke or dented wheel won't cause brake drag.
- *ABOUT TWICE THE BRAKE PAD AREA*
  — of Campagnolo and nearly all other side-pulls, for better fade resistance, more effective braking (wet or dry) and *much* longer life.
- *FULLY ADJUSTABLE SHOES*
  Ball and socket allows alignment in all directions.
- *MUCH GREATER PIVOT BEARING AREA*
  2-1/4 times as great as Campagnolo's, for longer wear, solider caliper arm mounting.
- *MUCH GREATER PIVOT BEARING LENGTH*
  3-1/2 times as great as Campagnolo's, for better fore-and-aft arm stability.
- *SWIVELING CABLE ANCHORING HARDWARE*
  For better cable core and casing alignment, lower friction, longer wear. All parts fully rotatable.
- *NEW CABLE CORE ANCHOR DESIGN*
  Easy insertion of frayed-end cable cores, and minimal cable deformation.
- *SPACE AGE MATERIALS*
  Ten parts per set are Titanium and Magesium, at no extra cost.
- *SHORTER CABLE ROUTING*
  Means lighter cables, less wind drag, less cable stretch (sponge) and neater appearance. No cables above handlebars. Invert bike without cable damage.
- *REPLACEABLE SYNTHETIC BEARINGS*
  To eliminate metal-to-metal friction at the arm pivots, and allow renewal when worn.

Expensive? Of course! Doesn't the world's best usually cost more?

Worth it? Certainly! Ask anyone who has a set.

And isn't it nice to see a top component made in the U.S. for a change?

---

dry conditions, of the the eighteen leading bicycle brakes from Italy and Japan plus one from the U.S.—the Scott Superbrake. Our brake had the best combined performance by a wide margin, proving once more that it isn't always necessary to go abroad for good design and workmanship. It is interesting that David Gordon Wilson extolled the beauty of the design. Both Scotty and I discovered independently that we now see the conventional Italian and Japanese brakes as quite ugly.

The variety of my work, if anything, increased as I approached my late seventies as the following partial listing proves: sailboard hardware for Navtec, leading manufacturer of high tech rigging; a patented, briefcase-size fire escape with which several persons can reach the ground from a twenty-story hotel window; a patented regenerative kitchen ventilator (this brought my U.S. patent total to ninety). In 1980 I prepared a long-range industry forecast for the Southern California Gas Company, for which I used my old friend and sailing companion Lee Moore to do most of the research. *Yachting* magazine had no one available to cover the finish of the 1980 Observer Single-Handed TransAtlantic Race in Newport so I was asked to do it. This was a special pleasure as one of my favorites, Phil Weld, aged sixty-five, was sailing a trimaran named "Moxie," designed by my old friend Dick Newick who then lived on Martha's Vineyard. Dick's years of struggle as a multi-hull protagonist finally paid off as Phil, by far the oldest skipper in the race, was the first to finish. I was at the helm of Phil's sixty-foot trimaran, Rogue Wave, as we towed the victorious Moxie into Newport's Goat Island Marina. It was a great experience. Everyone on the America's Cup Race Committee yacht lined the rail, doffed their white yachting caps, and bowed in unison in tribute to Phil's amazing feat.

I was writing as many articles as ever. One was the lead story in the *New York Times* Travel Section on hiking the various trails on the east shore of the Hudson, using the Hudson Line Railroad to get back

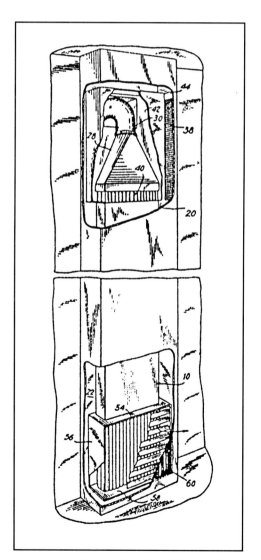
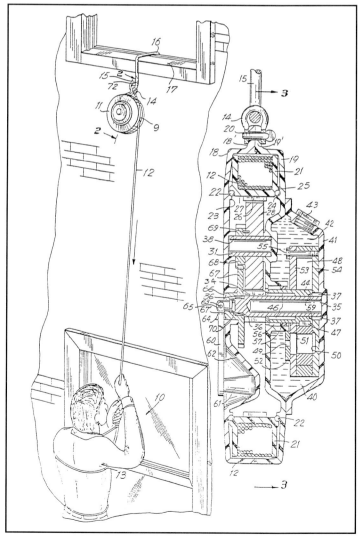

Patent drawings of Dorwin's regenerative kitchen ventilator and the briefcase-size fire escape.

Above: Dorwin as honorary guest at the "Marmon Muster," the annual meeting of the Marmon Owners Club at Boiling Springs, Pennsylvania in 1985. His bearded phase lasted several years. Below: A trimaran designed by Dick Newick. The third-generation Teague takes to the slopes; August (shown with Harry) on skis as early as his father had been and making a magazine cover too.

Snapshots from the sail to the Grenadine Islands, 1975, Dorwin and Harriette's only sail in the Caribbean that was not a bare-boat charter. The trip began in Antigua and wound up in Trinidad. In the B&B run by nuns there, the Teagues got into trouble one night for staying out too late.

Bicycling on Nantucket, Thanksgiving, 1979. The Lancia Scorpion, which replaced the string of Porsches in the Teague garage and stayed there a long time.

to the starting point. Others included an article on sailing in the fog for *Cruising World*, another on geriatric cruising that featured the various tricks we'd worked out for Soufflé to make short-handed sailing easier, and a third on the problems caused by changing aids to navigation. I did a story for *Small World*, the Volkswagen magazine, on bicycling on Nantucket over Thanksgiving. In March of 1984 Harriette and I chartered a J-30 in Key West and sailed to the Dry Tortugas and back, which was written up for *Yachting*. Other articles for *Yachting* were on sanitation devices and my own design for a 35-foot lightweight cruising trimaran. In 1985 Harriette and I started out on May 30th for a cruise to Baddeck in the Bras D'or Lakes in Nova Scotia, visiting sixty-three harbors and covering 2100 nautical miles before we got Soufflé back to her home port on the Hudson in early November. As a result of getting tangled with a lobster pot warp in Maine, I researched and wrote a long article for *Yachting* with special charts on the buoys and locations of fish traps and gill nets along the New England and Nova Scotia coasts.

In summer months we kept Soufflé on a mooring in Jamestown, Rhode Island, across the Bay from Newport but not yet gentrified or crowded. Harry's children were now old enough to enjoy cruising with us. The 1986 Liberty celebration in New York Harbor was something I would ordinarily avoid but I succumbed to son Harry's enthusiasm and we wound up anchored about a half-mile south of the Statue of Liberty surrounded by an estimated twenty to thirty thousand other boats. The catalyst, which made a happy adventure out of what could have been a miserable experience, was a pervasive spirit of good will among all the nearby boats. Every six hours as the tidal current turned we exchanged addresses and chatted with our neighbors as we put out fenders and untangled anchor rodes. As we sailed up to Rhode Island in easy stages, it was a great satisfaction to see how the third generation took to the cruising life.

Chapter Nineteen

1987 – 1998

"Dorwin, you're the one we want. Why can't you do it in the hospital?" ...

In May of 1987 I got a phone call from Tom Corcoran, ex-Olympic ski team member and head of a consortium of twenty-five ski area operators known as Dendrite Associates. They had banded together in 1980 to develop a more efficient snowmaking gun. Practically all ski areas everywhere make artificial snow to supplement natural snowfall. Even in the Rockies, with heavy annual snowfall, snowmaking has become imperative in order to groom trails and slopes.

There are two basic types of snowmaking guns—the so-called "airless" type which is cumbersome, expensive and difficult to move, and the "air/water" type, which is supplied with water and compressed air from hydrants scattered about the mountain. The latter is more flexible and less expensive, and is the dominant type. Its major drawback is the cost of compressing the air to a pressure of 80 to 100 psi.

Dendrite's objective was to develop a gun which could operate at lower pressures, hopefully at 30 psi. This could cut the energy cost of snowmaking by almost 50%. When one considers that the energy cost of snowmaking in one medium-sized ski area in the East is well above a million dollars a year, the 30 psi objective makes sense. Dendrite had retained the services of Calspan, an aero-space group, to study the basic technology of snowmaking and come up with a practical low pressure gun, if possible. Calspan had made its study and produced an experimental gun the performance of which, after seven years, was still unsatisfactory. Tom Corcoran asked if I would like to take a consulting contract to make it work. This was my idea of the ideal project. It combined my love of skiing with a development project that utilized my experience in fluid dynamics and mechanical design. I agreed at once.

Before starting to work, Harriette and I sailed Soufflé up to her summer mooring in Jamestown and took an Amtrak train back to Stamford where we had stashed a car in the municipal garage. Leaving Harriette at the station with our baggage, I was just stepping up to the curb by the garage when I was hit by a Nissan driven by a unlicensed female teenager. My left leg was in several pieces and I had wiped out her windshield with my head and shoulder.

There was no pain as I flew though the air or later as the EMS efficiently straightened out my leg and put me in the ambulance. The accident was completely unexpected—there was no other traffic or warning sound. I had heard that shock blanks out one's feelings; now I know it is true.

The next day I called Tom Corcoran from the Stamford Hospital and told him the bad news. I was due to be laid up for a long time and would be unable to do the job for Dendrite. Tom replied, "Dorwin, you're the one we want. Why don't you do it in the hospital?" I will forever be grateful to him. More effective therapy would be impossible to imagine. Within hours after the first operation, I was going over the Calspan reports and had a portable drafting board and calculator at my bedside. The first operation was botched. My leg had to be disassembled and strung on a steel tube à la shish-kebab some weeks later by my friend Dave Andrews, the top orthopedic man at Columbia Presbyterian in New York. Between operations I held a conference at my home in a wheelchair with half a dozen of the Dendrite leaders, who approved my proposal. The contract became official.

During this period my son Harry came east a couple of times to help set up meetings etc., and my friend John Lipscomb drove me around between hospitals in his family's comfortable minivan. By August I was walking with crutches and bought a Ford with a an automatic transmission so I could drive with one foot. Harriette and I attended a big Dendrite conference on Lake George and, later in October, Harry came east to help bring Soufflé back to the Hudson.

By this time I had witnessed field tests of the Calspan gun and had a pretty good idea of where I wanted to go. The gun had a number of rather obvious faults; the air inlet plumbing was tortuous and inadequate in size, the set-up was crude and the technology was based on a huge two-stage system which later proved undesirable. Snow could be made but not without a separate nucleator. Instrumentation was almost non-existent and no test records had ever been written.

Wally Shank, a young but highly experienced snowmaker, was assigned to help me. One of his assets was access to a competent and very reasonable machine shop in Gettysburg, Pennsylvania near his home. I designed a new gun which we called the Mark I and, by midnight of

Dorwin being tended after he wiped out his left leg and the Nissan's windshield. Photo: John Voorhees.

New Year's Eve 1987, we had completed a series of tests on two Mark I prototypes. The Mark I made snow even at comparatively warm temperatures with 30 psi air and no nucleator, and it began to appear that the second stage was doing more harm than good. In 1988, in the face of some opposition, I persuaded Dendrite to authorize a Mark II with double the output; tests showed an even better efficiency than the Mark I. The second stage was finally put to rest and on the basis of this performance, Dendrite started to think of selling the technology to an outside organization.

Toward the end of 1988 the Dendrite work was slowing down and the "freebie" I had done for the International Design Assistance Commission (IDAC) on the India Mark II water pump had come to life. Some of the ideas in my 1985 report were being considered and the number of Mark II pumps in the field was well over a million in developing countries of Africa and Asia. The World Bank, who was funding the deep well water pump project, thought it was time that I should get some first-hand experience with actual installations.

Accordingly, a guided tour was organized for me to visit Nairobi, Mombasa, Bombay, Ahmedabad, New Delhi and return via Muscat and London. By now I was getting around quite well and the idea of being paid to visit these exotic places was most welcome.

The success of the UN/World Bank deep well water pump program has been quite amazing, given its highly bureaucratic, committee-type development. The health of millions in the Third World population has been improved. Typically, each hand pump, which draws water from depths of 25 to 100 meters, provides relatively sanitary water for a village of 200 people. Even in slum districts of megatowns such as Delhi, the incidence of such horrible diseases as Guinea Worm is being eliminated. For me, it was especially rewarding to work on a project which was so beneficial to so many.

My gimpy leg was no great handicap on the trip since the World Bank provided efficient engineers and guides to lay on good transportation at each stop, minimize customs delays, and make sure I drank the proper water. In fact the only contretemps I remember was at a security check in one Bombay

This page: Dorwin's snapshots from the Africa trip. Page opposite: The India Mark II water pump, in widespread use in Mali. Photo: UNDP.

airport where the guard was giving each passenger a thorough going over with a hand metal detector. Each time he got close to my left leg with the 3/4-inch steel tube running from knee to ankle, the alarm sounded. Neither of us spoke the other's language. The whole security section was delighted when my sign language explanation was understood.

Actually the schedule was very tight in each country, which I personally prefer to leisurely traveling as a tourist. Villagers were invariably polite and cheerful, in spite of their extreme poverty. The schedule didn't prevent me from taking a couple of mini-safaris in the Nairobi National Game Park where we had many close-up views of wild game (our Land Rover was charged by a rhinoceros at one point) as well as swimming in the Indian Ocean and visiting the Taj Mahal.

Understandably, the technology in hand pumps is quite basic compared to rocket development but, on the other hand, it is complicated by the fact that before making any change, no matter how insignificant, one must consider the effect on the replacement parts for the one and one-half million pumps already in service. Also, national and regional politics play a major part in the overall efforts. A new development called the Afridev eliminates many of the original India Mark II's drawbacks but politics and the NIH factor (Not Invented Here) have virtually ruled out its acceptance in India. However, it is difficult to argue with the fact that there are now approaching two million India Mark II pumps in service and working, despite difficult maintenance and reliability problems. Development proceeds at the speed of a glacier, but the world is nevertheless better off than if the hand pump program had never taken place. It has been argued that saving another ten million lives a year by Third World programs like this is merely intensifying the world population explosion. Controlling world population is indeed our number one priority. But ignoring the horrible diseases that result from drinking impure water from polluted wells is not the way to do it. Actually a high infant death rate encourages a high birth rate, in hopes that one or two children may survive.

On my return from Africa and India I got to thinking about the eventual disposition of the Dendrite snowmaker which was being sold abroad—another example of a new technology developed to be demonstrably feasible and then handed over to Japan. I decided to see what would be involved in setting up a company to manufacture and market a snowmaking gun here in the U.S. The first obvious problem was financing such a venture. One of the most affluent members of the Dendrite associates was Bill Killebrew, whose Heavenly Valley near Lake Tahoe was the largest ski resort in the country. Because Bill had been actively encouraging the idea of building the snow guns here, I approached him. He showed interest, providing he could get the first 100 guns for his resort. Next I contacted Ed Scott, whose Scott USA had grown into a large organization before he sold out to a conglomerate.

I then went ahead on my own and made a layout of a production gun plus detail drawings of each part with all dimensions, tolerances, concentricities and material specs. On August 20th, 1989 the STK organization was formed. The initials stood for Snow Tech Kinetics (or Scott Teague Killebrew). Ed Scott was president, in charge of production, I was

United States Patent [19]

Teague

[11] Patent Number: **5,180,105**
[45] Date of Patent: **Jan. 19, 1993**

[54] SNOW MAKING APPARATUS

[75] Inventor: W. Dorwin Teague, 155 Tweed Blvd., Nyack, N.Y. 10960

[73] Assignee: Dorwin Teague, Nyack, N.Y.

[21] Appl. No.: 661,550

[22] Filed: Feb. 26, 1991

[51] Int. Cl.⁵ F25C 3/04
[52] U.S. Cl. 239/14.2; 239/407; 239/417.3
[58] Field of Search 239/14.2, 2.2, 405, 239/407, 417.3

[56] References Cited

U.S. PATENT DOCUMENTS

3,494,559 2/1970 Skinner 239/14.2
4,353,504 10/1982 Girardin et al. 239/14.2

Primary Examiner—Andres Kashnikow
Assistant Examiner—Lesley D. Morris
Attorney, Agent, or Firm—Fitzpatrick, Cella, Harper & Scinto

[57] ABSTRACT

An improved apparatus for producing snow is disclosed which includes a hollow water supply body having an integral water discharge throat formed therein at its front end and containing a needle valve assembly mounted at its rear end with an adjustable needle concentrically aligned with the orifice to control water flow through the orifice. A hollow air supply body is mounted on and receives the front end of the water supply body to receive water from the water supply body. The air supply body has an outlet orifice mounted thereon for discharging a mixture of air and water to a mixing tube for discharge to the atmosphere. Each of these components is clamped together in a single composite unitary construction with the bores and passage ways therein held in axially alignment by a plurality of clamping bolts.

23 Claims, 5 Drawing Sheets

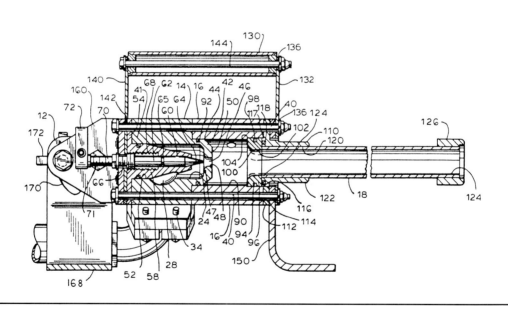

vice president in charge of design, Bill Killebrew was secretary and treasurer. Two prototypes of the production gun were built to my drawings and run for twelve hours on December 20th at Ski Liberty in Pennsylvania, side by side against a pair of the best competitive guns on the market. The STK guns were run with 30 psig air while the competition used the usual 80 or 90 psig. The results exceeded our expectations. The size of the snow pile in front of the STK guns was compared to a 747 fuselage, while the piles in front of the competition were pitiful in comparison.

After several months of bargaining, an agreement was signed between STK and Dendrite Associates on January 23rd, 1990 which gave STK exclusive world rights in the Dendrite "technology" based on its patent No. 4,916,911 in return for semi-annual payments. In addition to the drawings, helping set up production processes and supervising testing, I produced a detailed instruction manual and arranged to have it designed and printed. Meanwhile Ed Scott was lining up production sources and vendors for the purchased parts.

Scotty was doing a good job of organizing production but I began to worry that no one was handling any financial bookkeeping. Twenty pre-production guns were made up in time for the May Ski Industry Association Meeting in Boston but ski areas were suspicious of new developments and orders were slow. Scotty had taken over the financial management of STK from Bill Killebrew but was keeping no books or cash flow projections. Production was going well. Our target shipment slipped somewhat, but was still held to early November of 1990.

Testing was being conducted by Wally Shank at ski areas all over the country. Since he was getting a wet spot in some tests but not in others, Bill Killebrew had some question as to the performance reproducibility between guns. Because the U.S. ski season was over, it was decided to run tests in Australia, where Bill had some connections. The tests proved negative; there was no perceptible difference between one gun and another.

Another unnecessary, but costly, hassle involved the proposed distributor agreement with an outfit called Snomax, a subsidiary of Eastman Kodak. Its main product was a water additive that improved the performance of snowmaking guns, especially at marginally high temperatures. Snomax had representatives in most of the snowmaking countries, including Japan and all of Europe, seemingly an excellent opportunity for us. But Scotty took a dislike to Snomax and antagonized its president, John Gell. Scotty had made a handshake agreement with a Canadian company which Gell finally accepted although he didn't like it. For some reasons Dendrite got into the act at this point, although in my opinion, which I called to everyone's attention, the choice of a distributor, and the terms of our contract with him, was STK's prerogative, and required no endorsement from Dendrite. Three sets of attorneys—Snomax, Dendrite and STK—now entered into a typical lawyer's feeding frenzy, arguing back and forth over small points in the distributorship contract. Incredibly, this silly mix-up went on from April '90 to April '91, until finally Snomax became disgusted and pulled out. This not only cost STK a year of time, with huge payments to Dendrite when Snomax could have been an active distributor for STK, but the legal fees for STK's lawyers were accumulating and no one was paying any attention to their effect on the budget.

In addition, our Canadian distributor complained about the gun's performance. We finally discovered that he had been operating it with the air and water connections reversed, in spite of large engraved labels on each connection. The gun failed miserably before an invited group of Canadian ski area representatives, thus effectively killing any Canadian sales for the foreseeable future.

There were the usual minor problems, of course, attendant to any new production product. The synthetic nose guards fell off, some control knobs slipped, and there were three instances of the air inlet freezing at the elbow—all easily corrected. More serious was the fact that, when using high pressure (80 to 100 psi) air, common to all American installations so far, it was necessary to throttle the air at the hydrant to some extent, otherwise a wet spot would occur in the snow ahead of the gun. In some cases the wet spot didn't matter and, in others, the snowmakers quickly learned how to adjust the air. However, adjusting the air at the hydrant was alien to most snowmakers. Among the ski areas whose snowmakers were too lazy or too hidebound to learn how to operate the STK gun properly was Ski Sundown, whose owner was Dendrite Associates president Rick Carter. Rick organized a countrywide conference call among other disaffected members of Dendrite Associates to identify and investigate the good and bad points of the STK gun. Due to the make-up of the conferees, negative comments predominated. The widely circulated summary by Ski Sundown was quite effective at spreading the word that the STK gun was an "air hog" with lots of problems. Coming from the president of Dendrite, the negative opinions were accepted by many as gospel.

Even though the wet spot occurred only at high air pressures (for which the gun was not originally designed), we all knew that it would be far better if the wet spot didn't happen at any pressure. A couple of the STK group believed the problem was related to the fact that the air entered from the side, which caused the water jet to deflect, thus somehow causing the wet spot. The Dendrite group—including Mike Holden, inventor of the basic Dendrite patent—eagerly espoused this explanation. In order to determine whether or not the jet deflection was taking place, I suggested a "see-through" modified gun, but this was not done until four months later, when it showed no significant deflection. In the meantime, in case the deflection thing was valid, I designed a simple baffle which shielded the jet, without disturbing it otherwise. Made somewhat longer, my baffle resulted in a more efficient gun, probably because it caused some pressure drop which always improved the operation anyway.

One other variable that we had not thoroughly explored was the length of the so-called "mixing tube" or barrel of the gun. The long seventeen-inch barrel was a feature of Dendrite's No. 4,916,911

The Teague family. Son Harry (below) with wife Annie and children August and Emily, at MGM studio while visiting Lewis, 1990. Son Lewis (above right) with wife Elizabeth, 1998. Son Walter in that rarity, a good passport photo, 1997. Page opposite: Dorwin and Harriette, 1992. Teague wheels: Mitsubishi MSX. Having Thanksgiving at the Brick House, 1995: Lewis, sister Cecily, Cammy and her husband Norval White, Harriette. Photo by Dorwin Teague.

patent. Shorter barrels, it turned out, eliminated the "wet spot" problem completely, without affecting the rest of the performance. By this time, however, the combination of economic factors, negative performance propaganda, lack of a worldwide distributor, and absence of any cashflow supervision whatsoever, plus significant overproduction, resulted in the financial collapse of STK in June of 1991.

There was one encouraging development.

Frederick Hedin, a ski area owner/operator, whose resort is in southern Sweden, flew over to Australia in August of 1990 when STK was running performance tests and was given a gun to experiment with. Impressed with the way the gun performed, Fred decided to convert his entire area to 30 psi air and ordered twenty STK guns on the spot. His 1990–91 season was a big success and he saved 43% of his previous year's unit energy costs.

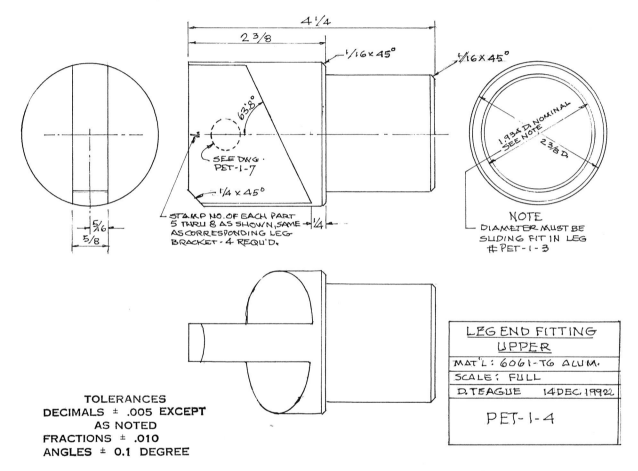

Page opposite: The Teague drawing for the mooring barge for Petersen's Boatyard. John Lipscomb at work in the shed. When finished, the barge was put in the water to see if it would float. It did. This page: The end of Soufflé. On the Mullica River, in the Pine Barrens, Palmer Langdon raises a bottle, and Dorwin poses next to a beaver dam.

This agrees closely with the theoretical energy savings for low pressure air. Since I now own the basic low pressure air patent outright, there is a unique opportunity for some organization with sufficient financing and marketing capability to make a killing in this expanding field.

At this writing I still seem to be in rather good health. Soufflé was wrecked in Hurricane "Bob," just after returning from a cruise to Naushon with my son Harry and his family. But I still sail occasionally with friends. A couple of years ago I arranged a fall cruise on the Hudson River for the Cruising Club of America of which I am a member. In my late eighties now, I still find that I'm happiest when solving an interesting design problem, to the point where I will take on a project without compensation if nothing else is available. One such was a super "mooring barge" for the ancient Petersen's Boatyard in Nyack, where I kept my boats for many years. My friend John Lipscomb and I were able to design and build a barge to lift moorings faster and more economically than ever before.

These days, a kayak replaces Soufflé, and I take an occasional canoe trip, such as a recent voyage down the Mullica River in the New Jersey Pine Barrens from Atsion to Batsto with friend Palmer Langdon—perhaps ten miles counting bends. Zero people all day. One can only hope the Pine Barrens will continue to resist population pressures.

Sadly, my dear wife Harriette died on January 7th of 1998 of complications from Alzheimer's disease. We had been friends from the ages of nine and eight. We had been married for fifty-nine years.

Epilogue

Some designers feel that anyone who is interested in the engineering aspects of new product development must be incapable of true aesthetic appreciation. This attitude lies partially in the history of the industrial design profession. Most early designers in America had backgrounds in stage design and advertising art. None had any engineering training except for Raymond Loewy who studied engineering for a short time in Paris, but never practiced it, and started his career as a fashion illustrator for *Vogue* and display designer for Macy's. Henry Dreyfuss and Norman Bel Geddes were stage designers, Russell Wright was a painter, and my father was an advertising artist. Academic industrial design courses were well in the future. The closest thing to them were architectural courses but these didn't offer the proper training for a product designer.

The term "industrial design" covers a wide spectrum of specialities, some of which require quite different skills than others. The early designers had to be versatile—in addition to product design, their output included display and exhibit design, interiors, packaging, graphics, corporate identity, architecture and anything else that was manufactured, printed or sold. This versatility was, still is, one of the most important ingredients of the industrial designer's repertoire, encouraging he/she to think more freely; be less inhibited about innovating and able to bring some things learned in one field to bear on others. The versatile designer is more likely to be able to forecast trends and to know what will most appeal to the customer. But, without some background in basic physics and engineering, no designer can consistently come up with acceptable product innovation. Like the early designers, he can only embellish the work of others.

Versatility doesn't mean a designer will be equally successful in all fields. Many fine graphic designers would be at a loss faced with a product design task and vice versa. The designer who thinks he can do all things equally well is fooling himself and will inevitably go down blind alleys and wind up with unhappy results. Explaining this fact to a designer who believes he is the exception is impossible; if he can't understand in his own mind that he is on a false track, it is hopeless to try to convince him of it. I remember one WDTA staff member who was working on a plasticene model for a ceramic ashtray with a miniature DC3 airplane on top of the dish part. Actually the DC3 was a practical, advanced design for the time but hardly an aesthetic breakthrough. Aesthetics aside, using it as an ornament on a ceramic ashtray was ridiculous, in the worst tradition of the schlock souvenir school. The plasticene model was made even more repulsive by giving the plane negative dihedral so that it looked somewhat like a sick water fowl expiring in a bird bath. I watched in awe as the designer worked on refining the object for almost a week, with all the panache and artistic temperament of a Michelangelo putting the finishing touch on the statue of David.

That someone very good in one design category might be a total wash-out in another was lost on my father. Although he always thought Bugattis were ugly, he could usually recognize good automobile design and felt that any well-qualified designer could design a good-looking car with, perhaps, a little guidance from himself. In 1953 WDTA was retained by the Packard Motor Car Company to come up with some new body design concepts. Packards have always been very highly regarded among classic car enthusiasts, both for their quality and their rather conservative, distinguished appearance. The car's unique hood and radiator contour, an unmistakable trademark, was difficult for others to copy, but also made it hard for the body designer to arrive at a sleek modern appearance. In difficult times by 1953, Packard had no inexpensive, high volume line to pay for the overhead and was slowly and inevitably heading for oblivion. A really outstanding body design, it was hoped, might help the company survive.

My relations with my father were at a low ebb at this time so, instead of the job coming to me, it went to Bob Harper. I think he wanted to prove to me, and to himself that my success with the Marmon Sixteen was nothing out of the ordinary, but rather a routine result which could be duplicated by any capable designer. Bob had excellent taste and was especially good with furniture design and interiors, but his knowledge of cars was minimal and his ideas for the Packard were pitiful. Finally I told my father that the Packard account was in trouble and that Harper's ideas were a disaster. "Well, why don't you help him out?" was his response. My father really believed that a nudge or two from me would produce a miracle. I did try to give Bob some suggestions but band-aid solutions were useless at this point and the whole account turned out to be a waste of time and Packard's money.

Automobile body design is primarily an aesthetic exercise, aside from rudimentary aerodynamic and ergonomic considerations. There is room for innovation but it must be evolutionary. The only successful body designers are people who grew up with automobiles and who instinctively know the direction of new trends. None of the early industrial design office heads had this kind of feeling and not one ever produced a really good-looking car. They could usually recognize good body design when they saw it but were incapable of creating it themselves. Loewy's Avanti, for example, was primarily the work of Bob Andrews who received no recognition for it. More typical of Loewy's own work were some of his early Studebakers with the projecting, pseudo-streamlined beak in the middle of the radiator. The really good early U.S. designs like Gordon Buehrig's Cord (1936) and the Bob Gregorie (with an assist from Edsel Ford) Lincoln Continental (1940) were the work of young designers who had lived and loved cars all their lives. As anyone can see at any major old car meet today, there have been many good car designs in the U.S. and Europe, all the work of automobile body designers. None, unless you want to admit my Marmon Sixteen, have been the work of industrial designers.

An aptitude for mechanical problem-solving plus my time at MIT allowed me to accomplish tasks like the line of vacuum cleaners for Montgomery Ward in 1939, which featured completely new principles and construction, but which were also aesthetically right. My experience at Bendix building up the Research Engineering Department to a profitable multi-million dollar operation convinced me that my future role should be in total development rather than in so-called styling. Later, the DTI office still did some of the more classic type of industrial design in which we were involved in the exterior casing of the product, including some automobile body design, but our main work was total development of new products. This requires more rigorous discipline; there is no place for a haphazard, pseudo-artistic approach in which the designer starts out by making a lot of sketches in the hope that one of them will turn out to be the perfect answer. We insisted that the problem should be thoroughly defined before getting locked in on any specific design or direction. The formal procedure we used, which was very successful, is described in Appendix A.

Aesthetics and mechanical aptitude are not mutually exclusive. Leonardo is the classic example of how the two go hand in hand, and I have cited a number of artist/engineers of my own acquaintance. Actually the whole profession of industrial design is undergoing something of a metamorphosis in this direction with regard to product design. My own interest in total development has rubbed off on a number of my fellow workers. Ben Stansbury, who worked on the Ritter dental chair, subsequently had

Dorwin Teague, summer of '98. Photo: James H. Cox.

a thriving experimental shop in California, where he also developed an unsuspected aptitude for politics and was twice acting mayor of Beverly Hills. Gary Van Duersen was head of Black & Decker's European operation and has several patents on new devices to his credit. Tip Sempliner has a development shop in Douglaston, Long Island, where he has designed a number of innovative products including some very effective streamlined bobsleds for U.S. team members. Wes Verkaart, as mentioned earlier, started his own medical equipment design and manufacturing company which continues to grow and prosper. Others, too numerous to mention, have found, as I did, that free-wheeling industrial design training, in which the designer constantly jumps from one field to another, provides a decided advantage over the engineer who has concentrated on one special category during his whole career.

Nearly all engineers today are specialists. As our technology becomes exponentially more complex, the scope of each engineer's responsibility becomes narrower and more concentrated. The more versatile engineer types are likely to become department heads and, unless they are careful, spend most of their time in meetings and on paperwork and administrative duties, which limits their involvement in the actual design process. The industrial designer, especially the independent consultant, is practically the only non-specialist left; he is less inhibited in his thinking and more able to see the forest as well as the trees. Obviously, the industrial designer can't replace the specialist engineer but there are many opportunities for him to bring a fresh approach to bear, which occasionally turns out to be a very important contribution. Who would have expected that an industrial designer like myself would have been selected to prepare a sizeable twenty-year forecast for the gas industry with recommendations for action in marketing, supply and future research?

As head of the Research Engineering Department at Bendix, I made it a practice to spend a substantial amount of my time on the actual mechanics of new product development. I did a lot of my own calculations, which I could have assigned to others, and kept my own drafting board where I made careful drawings of new concepts, not only because I had an instinctive fear of losing touch with the creative process, but also because I enjoyed it. I was unique among department heads in this respect but felt it was important to strike the proper balance between administrative duties and the nitty-gritty aspects of design and development. There is a potentially valuable opportunity for training industrial designers, or product designers, also non-specialist engineer types. A lot remains to be learned about the proper amount of training in important engineering fundamentals that will still keep the relatively fragile open minded quality alive in promising pupils.

In this book I've tried to emphasize the importance of outside interests—sports and cultural activities far removed from my vocation. Age hasn't diminished my enthusiasm for sailing, hiking, skiing—if anything, I spend more time on these sports than I did when I was younger. The longest cruise Harriette and I ever made was in 1985 when, in our mid-seventies, we took the whole summer and sailed Soufflé to the far end of Nova Scotia and back. I've even added a new spring and fall activity; participation in the local hawk migration count on the top of a nearby hill, where we've had over 14,000 hawks on one day. My only regret is that my conversion from bird shooter to bird watcher didn't take place while my mother was alive. Music is still important; every once in a while I collect a new Ellington record and I trade tapes back and forth with my architect son.

I've never been a very good Christian. If I had to choose among the formal religions I'd probably pick the old Shinto faith. Shintoists worshipped nature and held that man was only one of its component parts—animals, trees, even rocks were just as important. Christian faith teaches that man is supreme over nature and all other creatures. Actually, through overpopulation and the resulting destruction of forests and wildlife, man is inevitably destroying the world and himself. My spiritual experiences are the sudden view of New Hampshire and Vermont as the fog lifts on top of the Presidential range; a moonlight sail in the Caribbean tradewind; skiing alone in untracked powder in a pine forest; watching a peregrine dive at terminal velocity on a decoy owl; listening to a solo passage of music by Johnny Hodges; the feeling that comes when I realize that I've got a perfect solution to a design problem. This philosophy may seem paradoxical for one whose profession is the development of high tech gadgets, but actually the two are not incompatible.

Our priorities have been all wrong for too long. Only recently has the environment been given any serious consideration. When the day of reckoning comes, as it surely will, we'll need all the talent we have available to devise benign, renewable energy sources, to figure out drastic conservation measures, to rehabilitate mass transportation systems, and all the other duties we have been sweeping under the rug. This will present a special opportunity for industrial designers, the artist/engineer types, who have seen this crisis building up and have been thinking about ways to mitigate the inevitable trauma of revising our basic goals and our way of life. The physicist, engineer or chemist who has been spending a good part of his career on space-based weapons of destruction or biological warfare will need a while to adjust to completely new peacetime activities. We will be faced with the formative stages of many new endeavors, where broad-based thinking will be required and the non-specialists will come into their own.

One of the rewards of writing this book was the discovery of the autobiography of my freshman calculus professor, Norbert Wiener. His writing came as a surprise and delight to me; his philosophy is so close to my own and he expresses it so well. His experience in writing his autobiography was very similar to my own and my best course is to quote him directly:

"There is a problem which must be faced in one form or other by an autobiographer who has done significant work in a difficult and private field like mathematics. A composer cannot avoid paying a certain amount of attention to the techniques of composition, and to aspects of harmony and counterpoint which are the very substance of his work but which can be appreciated only in a very limited degree even by the devoted concert goer who has not himself faced the task of musical composition. The writer or the painter is no less involved in the problem if he goes in for autobiography. He may seem to address himself to be an educated layman who can appreciate the results of his creative work. However, he has not completely performed the task of the autobiographer unless at the same time he has managed to express himself concerning the tasks of writing and of painting which can be fully appreciated only by the man who himself has faced them on a high professional level."

In industrial design, as in mathematics and all other professions, integrity is a vital ingredient. The ideal professional combines ability, integrity and salesmanship but, as we know, the three don't always go together. Just as a politician may get by, or even develop landslide support, on nothing but personal charisma, so an industrial designer can be successful, at least for a while, without integrity or even much ability if he has sufficient charm and persuasiveness. Although such people are eventually exposed, they may achieve considerable popular acclaim or financial success until that happens. We all know examples of ugly design trends which were huge marketing successes in their time, such as the monstrous tail fins, chrome plate and excessive size of American cars in the 1950s. Pursuit of such useless ostentation became so ingrained in the Detroit consciousness that European and Japanese makers were able to grab a substantial share of the market before Detroit could change its thinking.

The safe, easy road for a designer is to accept current trends and styling clichés, whether or not they make any practical sense. This way no enemies will be made but nothing of lasting significance will be accomplished either. The really good designer will question such clichés. If he knows in his heart that they are superfluous or unwarranted, he will resist the easy road of going along. This may cost him an occasional account but, in the long run, he will be better off. The opposite course, to change something merely for the sake of being different, when the change is not as desirable practically or aesthetically as the old way of doing it, is just as bad as slavishly following trends. In such fields as ladies clothing, eyeglass frames, expensive watches and furniture, the designers compete with each other to see who can come up with the most outlandish styling, regardless of practicality. This doesn't do any particular harm, it can even be temporarily amusing, but it insures early oblivion for the design.

If a new product design looks quite different from anything of its type in the past, and this results in real improvement in function or in ease of manufacturing, then the difference is justified. Sometimes it may take quite a while for the general public to get used to the new design but if the new concept is really and truly superior, it will look right to the designer and eventually to everyone. Occasionally, if we're lucky, we may come up with a new design for a product which is so obviously right that everyone recognizes its beauty at first glance. Those are the times when we're glad we're in the design business.

Appendices

APPENDIX A
Dorwin Teague, Inc. Design Process 221

APPENDIX B
United States Patents 224

APPENDIX C
Life List of Projects 228

APPENDIX D
"Resourceful Design Thinking Behind a New Cleaner Group," *Electrical Manufacturing*, December 1940 233

APPENDIX E
Report to Edsel Ford Recommending a Ford Sports Car 239

APPENDIX F
"The Twenty Best Industrial Designs Since World War II," *Saturday Review*, May 23, 1964 245

Appendix A

Dorwin Teague, Inc. Design Process

DTI developed a technique for dealing with design problems that was new and appeared to be highly successful. As time went on, we began to think of ways that we could formalize the approach a little, and train our people to take advantage of it. We have applied the technique to all kinds of jobs—buildings, small one-piece plastic items, packages, large complex items and complete systems. The process as finally developed became so effective that DTI was hired by large organizations such as Honeywell to indoctrinate their designers and engineers in how it works.

In a way, what we have is a simplified, non-computerized version of a formal systems analysis or Operations Research System. We worked it out completely independently but it is gratifying that there are so many similarities.

The following is a description of how the technique is conducted.

1. IDEAL OBJECTIVES—The first step is always the same; the preparation of the major or "ideal" objectives. This is done before any design at all, before "researching" or anything else. This step consists of writing down a series of short, concise statements on "what we would like to do in this design."

This sounds simple until you actually try it; it is much harder than it sounds.

The technique is important. Each objective is argued over and discussed until it correctly represents the formal consensus of everyone concerned with the project.

Basically, the objectives are the ultimate or optimum design goals. When formulating these one should not worry at all whether or not these objectives seem practical or even attainable. The designers should let themselves go. In other words, if the sky is the limit, what would we like to see in the final results? Also we may think of objectives that are conflicting. No matter, put them down anyway.

We start with a preliminary list of objectives and, as the job progresses, we add to or modify the list. At the time the objectives are put down, some will seem pretty impossible to meet, but a lot of good opportunities would be missed if we didn't set our sights as high as possible.

As in formal systems analysis, each objective is now rated as to relative importance in the overall picture. The scale is one to ten, with ten being the most important or "least negotiable." Such items as safety features might rate a ten.

One important result of this preliminary stage is that everyone is thinking along the same lines as to what we are trying to accomplish. The client as well as the design team should participate in the final selection of objectives. In this way we make sure there are no misunderstandings later on.

2. IDEAS—Working on designs or engineering details is specifically forbidden until the objectives are listed but, of course, you can't prevent the participants from thinking of them. As the objectives are being prepared and discussed, solutions will inevitably occur to the creative type working on the job. These are all written down and filed for future reference. No matter how screwy some of these ideas may sound, they are still listed. Later on they are considered during the preliminary design phase. Also, at this point, one should avoid examining competitors' solutions to the problem. These will be considered later, but concentrating on existing methods will inhibit new ideas and almost insure a "me too" result.

3. PRELIMINARY DESIGN SOLUTIONS—This is the part that everyone has been waiting for. By holding off design work until this point, you have probably got an engineering/design team that is about ready to burst with

Table 1

Basic objectives of window radiation control system.	Rating *
1. Prevent exterior radiation, direct and indirect, from entering room interior when desired.	10
2. Permit all radiation to enter into room when desired.	9
3. Prevent interior heat from being re-radiated at night when desired.	7
4. Permit interior heat to be re-radiated at night when desired.	4
5. Provide maximum degree of insulation when desired.	6
6. Present minimum obstruction to view at all times.	9
7. Perform all functions with minimum manual control.	8
8. Be as easy as possible to clean.	6
9. Be as inexpensive as possible.	9
10. Require minimum maintenance.	9
11. Be as architecturally attractive as possible.	8

* Rating indicates relative importance on scale of one to ten, ten being most important.

new ideas for the product. Don't worry about what others have done, don't worry about cost or availability of materials. Try to solve the objectives in the ideal manner. We start referring to our list of crazy ideas, and we usually add a lot of new crazy ideas. We don't rule out any solution because the X company tried it in 1963 and it was a big flop. What we try to do is to design the best way from the functional standpoint and to satisfy the objectives.

All possible solutions to the problem should be included. Regardless of how implausible or far-fetched, no solution is excluded at this stage. Once all our new ideas are listed, we can examine what others have done and add these to the list.

The next step is to take each one of those ideas or solutions and rate them for degree of compliance with each objective. One solution will satisfy some objectives but will not comply well with others. No solution ever gives a perfect answer for all objectives. Again, we rate each solution for degree of compliance on a scale of one to ten with each objective.

Finally, we multiply the degree of compliance by the importance rating of each objective for the various solutions, old and new, and the sum of these products represents the "score" for each solution. The solution with the highest score is the way to go.

Now this sounds a little bit like we are advocating design by the numbers, which would suggest that an accountant might do a better job than a professional designer. Obviously this is not so. The main purpose of the system is to substitute thinking for haphazard sketching, and to make sure that no really good solutions are overlooked. This doesn't rule out inspiration and intuition; on the contrary it avoids getting locked into mediocre or "me too" concepts before all possibilities are explored. It may even turn out that the old conventional way is the best but, in the majority of cases, I have been able to come up with new and patentable better approaches and to convince the client that these are the way to go.

As an example of how the Ideal Objective System works, Tables 1 and 2 show the objectives and the comparison with other types of window control for the Automatic Electric Window Radiation Control which I later proposed to Polaroid.

Table 1 gives the relative importance of the various objectives on a scale of one to ten, with ten being the most important. Table 2 gives a compilation of the compliance of various other window radiation control methods with the objectives. The left-hand figures in each column give the degree of compliance with each objective while the right hand figures are the degree of compliance times the objective rating. The totals at the bottom are the sum of the products and give an indication of the relative desirability of each system. The automatic electric control is obviously the best of the lot in spite of a poor cost rating (objective number 9). This brings out an inherent limitation in the rating system since "cost," which normally conflicts with the other objectives, can be a more serious sales deterrent than the low rating suggests. For example, if the electric control system were to cost $100 per square foot, the market would be very limited. In this case, however, the cost is expected to be low enough to allow a substantial market.

Later on certain additional objectives were added such as the desirability of do-it-yourself installation and a manual over-ride to allow momentary outward vision or to completely opaque the window from outside viewing for privacy. It is important to modify the objectives and their ratings as the project progresses.

Table 2

Objective Number	Objective Rating "R"	Bead Wall (R × DOC)		Trellis or Awning		Interior Venetian Blinds		Interior Vertical Blinds		Fixed Exterior Horizontal Louvers		Manually Adjustable Horizontal Exterior Louvers		Sandwich Venetian Blinds		Electric Window Radiation Control	
1	10	10	100	6	60	3	30	3	30	7	70	10	100	9	90	9	90
2	9	10	90	7	63	9	81	9	81	7	63	9	81	9	81	9	81
3	7	10	70	3	21	8	56	8	56	4	28	8	56	10	70	9	63
4	4	10	40	6	24	8	32	8	32	6	24	8	32	8	32	9	36
5	6	9	54	2	12	1	6	1	6	0	0	3	18	7	42	7	56
6	9	5	45	9	81	7	63	2	18	7	63	7	63	7	63	10	90
7	8	7	56	10	80	4	32	4	32	10	80	9	72	4	32	10	80
8	6	5	30	10	60	5	30	2	12	3	18	4	32	8	48	10	60
9	9	1	9	8	72	8	72	7	63	7	63	5	45	7	63	1	9
10	9	1	9	10	90	8	72	8	72	9	81	6	54	8	72	10	90
11	8	6	48	8	64	7	56	8	64	3	24	4	32	8	64	10	80
TOTAL SCORE			551		627		530		586		514		585		657		735

Overall Ratings of Window Radiation Control Systems

Appendix B

Dorwin Teague's United States Patents
(foreign patents not listed)

Number	Filed	Description	Assignee
107461des	1936	Coal/gas range	Floyd Wells Company
113900des	1938	Cash register	National Cash Register Company
115177des	1938	Rear-engined car	Inventor
124567des	1940	Extension leaf table	Enamel Products Company
124657des	1940	Extension leaf table	" " "
124658des	1940	Extension leaf table	" " "
186914des	1959	Ice bucket	Columbus Plastic Products, Inc.
187864des	1959	Dental chair	The Ritter Company
188864des	1960	Dish dryer	Columbia Plastic Products
205263des	1965	Smooth Top range	Columbia Gas Company
205264des	1965	Cooking oven	" " "
236328des	1974	Bicycle reflector	Amerace Company
236329des	1974	Bicycle reflector	" " "
2,100,293	1936	Gas range cover	Floyd Wells Company
2,152,106	1937	Radio dial	Crosley Corporation
2,221,063	1937	Cash register read out	National Cash Register Company
2,295,094	1940	Extension leaf table	Enamel Products Company
2,295,095	1940	Extension leaf table	" " "
2,295,096	1940	Extension leaf table	" " "
2,362,948	1943	Self closing pen	Schaeffer Pen Company
23,445C1	1944	Cabin pressure valve	Bendix Aviation Corporation
2,411,816	1944	Centrifugal blower	" " "
2,428,593	1941	Telephone hand set	Kellogg Switchboard
2,441,088	1944	Cabin pressure valve	Bendix Aviation Corporation
2,441,089	1944	Cabin pressure regulator	" " "
2,459,200	1943	Meat slicing machine	U. S. Slicing Machine Company
2,473,620	1944	Suction relief valve	Bendix Aviation Corporation
2,488,647	1944	Pressure throttle valve	" " "
2,488,648	1944	Suction regulator valve	" " "
2,488,649	1944	Suction regulator valve	" " "
2,490,174	1947	Servo actuator	" " "
2,507,945	1944	Supercharger control	" " "
2,542,167	1947	Position control motor	" " "
2,556,829	1945	Fluid regulator	" " "
2,561,957	1947	Hydraulic accumulator	" " "
2,573,231	1947	Piloted relief valve	" " "
2,600,137	1947	Pressure regulator	" " "
2,612,757	1947	Dual source turbine	" " "
2,628,594	1947	Electric/Hydro servo	" " "
2,638,107	1948	Suction regulator valve	" " "
2,639,722	1947	Pump unloading valve	" " "
2,642,543	1951	Emergency power unit	" " "
2,647,730	1949	Oil cooler protector	" " "
2,664,710	1949	Steam starter	" " "
2,666,449	1949	Dual pressure control valve	" " "
2,700,935	1948	Rocket fuel pump	" " "
2,720,282	1951	Oil cooler bypass valve	" " "

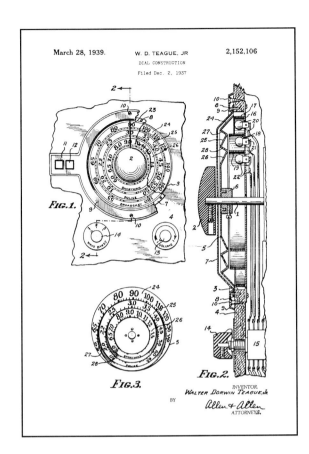
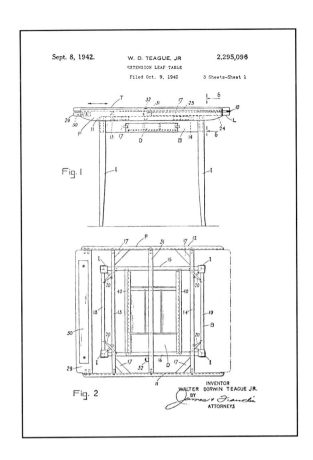
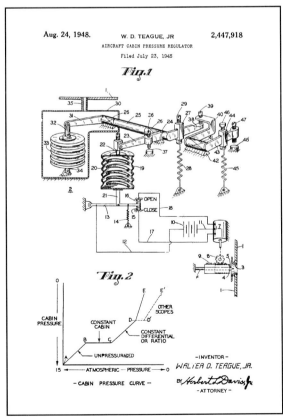
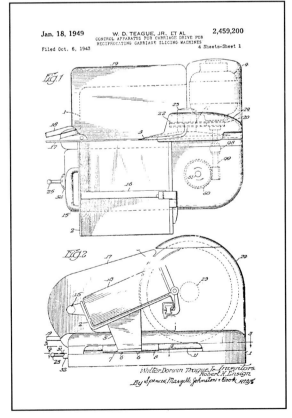

2,731,526	1951	Acceleration switch	Bendix Aviation Authority
2,736,167	1950	Ram-jet fuel ratio control	" " "
2,829,491	1954	Two-stage liquid rocket	" " "
2,841,953	1955	Rocket pressurizer	" " "
2,850,975	1954	Acceleration rocket	" " "
2,859,768	1953	Fluid flow regulator	" " "
2,896,919	1959	Oil cooler by-pass valve	" " "
2,939,017	1949	Air-driven power supply	" " "
2,940,332	1955	Ratio changer	" " "
2,940,696	1955	Differential control mixer	" " "
3,010,761	1958	Reclining chair mechanism	Owned by inventor
3,025,108	1959	Dental chair	Ritter Dental Company
3,027,041	1960	Dish drainer	Columbus Plastic Products
3,165,801	1962	Ropehold	Hartford Company
3,209,124	1963	Humidifier	Keeney Manufacturing Company
3,267,540	1965	Cord fastener	Hartford Company
3,302,229	1964	Ship's gangplank	Puerto Rico Parts Authority
3,407,726	1968	Manual printer	Dorwin Teague, Inc.
3,451,650	1967	Magnetic shut off	Columbia Gas System
3,468,298	1967	Smooth-top gas range	" " "
3,779,831	1967	Pipe-laying system	" " "
3,514,078	1967	Gas range control valve	" " "
3,537,483	1967	Retracting pipe plug	" " "
3,547,507	1968	Pull-out cabinet	Tappan Company
3,572,314	1968	Heated diver's suit	Columbia Gas System
3,668,784	1970	Clothes dryer	" " "
3,696,802	1970	Compact gas heater	Southern California Gas Company
3,734,363	1971	Boot jack	Dorwin Teague, Inc.
3,771,945	1972	Infra-red gas burner	Southern California Gas Company
3,885,907	1972	Modulating infra-red burner	Columbia Gas System
3,934,572	1972	Infra-red space heater	Dorwin Teague, Inc.
3,944,237	1974	Ski binding	
4,175,299	1977	Water power toothbrush	Sterling Drug Company
4,175,359	1977	Water power sander	" " "
4,179,765	1978	Nutating motor	" " "
4,210,975	1978	Faucet connector	" " "
4,229,150	1978	Anti-rotation device	" " "
4,276,672	1977	Water power toothbrush	" " "
4,281,681	1977	Diverter valve	" " "
4,336,622	1981	Water power toothbrush	" " "
4,353,141	1980	Water power toothbrush	" " "
4,437,545	1981	Fire escape device	Gerald Marinoff
4,497,362	1983	Regenerative air exchanger	Southern California Gas Company
4,553,641	1985	Bicycle brake	Dorwin Teague, Inc.
5,180,105	1992	Snowmaking gun	" " "

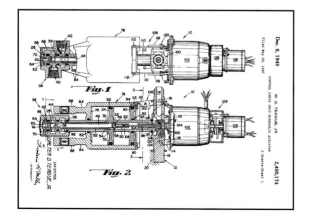

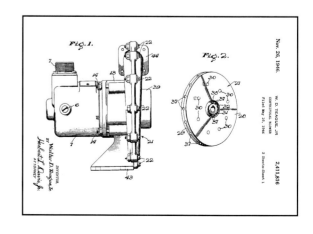

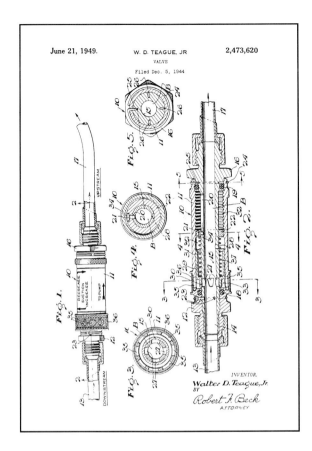

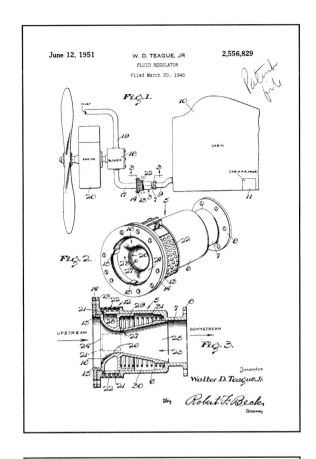

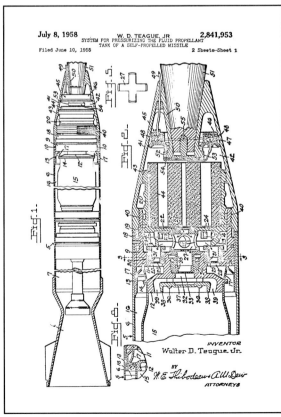

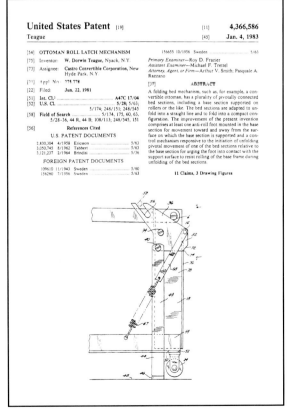

Appendix C

Life List of Projects by W. Dorwin Teague

Year	Project	Client
1927	Do-it-yourself ashtrays, etc.	Popular Mechanics
1929	Marmon Sixteen automobile body design	Marmon Motor Car Company
1931	Automobile model	self
1932	Sculpture of lion	Steuben Division of Corning
	Marmon Twelve automobile body design	Marmon Motor Car Company
1933	Oil burning stove	American Gas Machine
	Gas stove	Floyd Wells
1934	Humidifier	Bryant Heater
	Pressure regulator	" "
	Dualator	" "
	Gas Heater	" "
	Dehumidifier	" "
	Conversion burner	" "
	Warm air furnace	" "
	Boiler	" "
	600 HP Radial Gas Engine	Dresser Manufacturing
1935	Mimeograph Machine	A. B. Dick
	Tractor	Caterpillar Tractor
	Model 100 Cash Register	National Cash Register
	Gas stove	American Gas Machine
	Gasoline pump	Neptune Meter
	Gas ranges	Floyd Wells
1936	Warm air furnace	Bryant Heater
	Conversion burner	" "
	Unit heater	" "
	Radio cabinets	Crosley Corporation
	Xervac machine	" "
	Air conditioner	" "
	Mimeograph	A. B. Dick
	Glass brick	Pittsburgh Plate Glass
	Lantern, gas stove	American Gas Machine
	Display stand	Ferry Morse Seed
1937	Oil burning space heater	Crosley Corporation
	Packaging graphics	A. B. Dick
	Display railing	Ford Motor Company
	New York World's Fair exhibit turntable	Ford Motor Company
	Cash registers	National Cash Register
	Gasoline truck	Texaco
	Air-conditioning unit	York Ice Machine
1938	Street lighting	National Cash Register
	Internal grinder	Heald Machine Tool Company
	New York World's Fair trains	New York World's Fair
	Glass brick	Owens Illinois Gas
	Space heater	American Gas Machine
	Antique car models	Ford Motor Company
1939	Mimeograph	A. B. Dick
	Facsimile unit	Finch Telecommunications
	Gas boilers	Bryant Heater
	Furnaces	" "
	Circulator	" "
	Bicycle	Montgomery Ward Company
	Vacuum cleaners	" " "
	Carpet sweepers	" " "

	Boilers	Montgomery Ward Company
	Oil burner	" " "
	Lamp shade	Polaroid Corporation
1940	Sports car design	Ford Motor Company
	Breakfast sets	Enamel Products Company
	Furnaces, boilers, air conditioner	Bryant Heater Company
1941	Sports car design	Ford Motor Company
	Strategic equipment	OSS
	Sleds	S. L. Allen Company
	Breakfast sets	Enamel Products Company
	Telephone set	Kellogg Switchboard Company
	Desk pen sets	Schaeffer Pen Company
	Slicing machine	U. S. Slicing Machine Company
	Truck body	White Motor Company
1942 through 1952	Eclipse Pioneer Division of Bendix Aviation Corporation, Junior Engineering in Research Engineering Department. In 1945 was made head of Research Engineering Department. Products for which I was responsible included the following:	

Modifications to B17 cabin supercharger
Cabin pressure regulators
Steam starter for jet fighters
Instrument throttling valves
Oil cooler emergency by-pass valve
Three-stage high pressure cabin supercharger
Lark surface to surface liquid propellant rocket
Self-contained emergency power supply
Loki liquid propellant anti-aircraft rocket
Hydrogen peroxide jet engine starter
Fuel-air jet engine starter (production factory built in Utica, New York)
Thirty-three patents were taken out in my name.

1953	Returned to Walter Dorwin Teague Associates as Associate Partner, made full partner in 1954.	
1953	Nylon .22 rifle	Remington Arms Company
	Folding door	Columbia Mills, Inc.
	Variable ratio linkage	Bendix Aviation Corporation
	Piano & bench	Steinway and Sons
	Inspection process	U. S. Navy, Board of Ordnance
1954	Barcaloafer chair	Barcalo Manufacturing Corporation
1955	Brick handling truck	Ohio Clay Company
	Engineering consultation—Sweden, Norway, France, Switzerland	Servel Corporation
	Autoclave	Ritter Company
	Navy inspection application	U. S. Navy Bureau of Ordinance
1956	Design building, Gorky Park, Moscow	U. S. Department of Commerce
	Prototype Barcaloafer	Barcalo Manufacturing Corporation
	Telescoping mast	H. K. Lorentzen Inc.
	Dental chair	Ritter Company
1957	Exhibit building—Zagreb	U. S. Department of Commerce
	Exhibit building—Vienna	" " " "
	Folding door	Columbia Mills, Inc.
	Install Idlewild model, Brussels	Port of Authority of New York
1958	Atoms for Peace exhibit, Geneva	Atomic Energy Comm.
	Ice bucket	W. R. Grace Company
	Dish drainer	" " " "
	New erecting system	Polaroid Company
	FAS mobile exhibit	Reynolds Feal Company
1959	Amsterdam exhibit	American Horticultural Council
	Mobile airport lounge	Chrysler Corporation
	Barcelona trailer exhibit	Department of Agriculture
	School furniture	W. R. Grace
	Fertilizer spreader	" " "
	Shipping cart	" " "
	Information Center	Longwood Gardens
	Advanced cash registers	National Cash Register
	Civil War Information Center Building	Virginia Civil War Commission
	Bedroom furniture	Commonwealth of Puerto Rico

Year	Project	Client
1960	Gas Building, 1964 New York World's Fair	Gas Inc.
	Plastic Cooler	W. R. Grace
	Treadle garbage can	The Hartford Company
	Humidifier	Keeney Manufacturing Company
	Traffic flow study	Boeing
1961	Gas Inc. building model	Gas Inc.
	Ropehold device	The Hartford Company
	New Jersey Historymobile	N. J. Tercentenary Commission
	Gang Plank System	Puerto Rico Ports Authority
1962	Gas Building Interiors	Gas Inc.
	Plastic product concepts	W. R. Grace
	Humidifier	Keeney Manufacturing Company
	Conveyor belt system	Puerto Rico Ports Authority
1963	Building construction	Gas Inc.
	AIAA symbol (still in use 1998)	American Institute of Aeronautics & Astronautics
	Ratio changer mechanism	Bendix Aviation
	Restaurant	Gas Inc.
1964	Modular exhibit system	Department of Commerce
	Humidifier	Keeney Manufacturing Company
	Install Idlewild airport model, Brussels	Port Authority of New York
1965	Trade show exhibits, Berlin, Holland, Germany, Sweden	U.S. Department of Commerce
	Smooth-top range concept	Columbia Gas Company
	Smoke detector	Fenwal
	Petersberg Civil War Centennial Building	Petersberg, Virginia
	Clothing merchandising system	Klopman Mills
	Packaging in Jamaica	USAID
1966	Resigned from WDTA, started Dorwin Teague, Inc. in Englewood Cliffs, New Jersey	
	Underwater habitat, Hawaii	Columbia Gas System
	Inspect Gulf oil rigs	" " "
	Underwater pipe laying system	" " "
	Smooth top range working model	" " "
1967	Infra-red burner	" " "
	Advanced clothes dryer	" " "
	Servo valve	Moog Inc.
	Corporate identity	" "
	Electric broiler	Nautilus Industries
	Production smooth top	Tappan Company
	Advanced kitchen system	" "
1968	Milling machine tool head	Moog Inc.
	Tappan/Thermo Electron exhibit	Tappan Company
	Cold storage system	" "
	Display racks	O. M. Scott
	Infra-red burner	Columbia Gas System
1969	Advanced clothes dryer	" " "
	Advanced refrigeration study	" " "
	Gas snow removal system	" " "
	Design method lecture	Honeywell
	Fire alarm system	"
	Advanced space heater	SOCAL (Southern California Gas Company)
	Mobile home gas appliances	" " " " "
	Proportional control valve	Unitec
1970	Solid state programmer	Amerace/ESNA
	Cryogenic tank study	Columbia Gas System
	Cryogenic gas terminal	" " "
	Design indoctrination	Manitoba Province
	Magazine cover	Motor Boating Magazine
	Ski boot hardware	Scott, USA
	Glove box	Scott, USA
	Broiler evaluation	SOCAL
	Formula Vee logo	Volkswagen of America
	Headrest system	" " "
	Auto Union logo	" " "
	People heater	Unitec

	Heat shrinker	Unitec
1971	Camping equipment	Garcia Corporation
	Burner market study	Columbia Gas System
	Porsche 914 body design	Porsche of America
	Minical heater	SOCAL
	Institutional hot water heater	"
	Roll-up fence	Willow & Reed
	Commercial griddle	Thermo Electron
1972	Wire nut designs	Buchanan Electric Company
	Terminal boards	" " "
	Ski binding	Moog Inc.
	Non-conductive data link	"
	Burt ski binding	National Recreational Industries
	Night club	Steamboat Springs
	TV projector	Vior Corporation
1973	Domestic central services system	Consolidated Gas Company
	" " " "	Lone Star Gas Company
	" " " "	Northern Illinois Gas Company
	" " " "	SOCAL
	Agastat relay	Amerace
	Auditory trainer	Electric Futures, Inc.
	Portable controller	Moog
	Reception area	"
	Sphygmomanometer test report	Patient Care Magazine
	Cyclonair blower	Rotron
	Infra-red heater license	Sears
	Ski binding test rig	"
	12 volt Minical heater	SOCAL
	Ski binding design	Stanley Tool Works
1974	Bicycle reflector kits	Amerace
	Nautical instruments	Occidental Petroleum
	Minimum Energy House proposal	SOCAL
1975	Expert witness Phillips v. Dictaphone	Dictaphone
	Expert witness Motorola v. Lear Jet	Lear Jet
	Carbon fiber hardware	Keithley Molding
	Helio Kinetic Simulator	SOCAL
	Molded bicycle brake	Scott/Mathauser
	50th Anniversary package	Leitz Inc.
1976	Carbon fiber winch handle	Keithley Molding
	Nutating water motor	Sterling Drug Company
1977	Cigarette-making machine	Pyro Dynamics
	Clothes dryer	Clairol
	Ski binding	Von Besser
	Solar collector prototype	SOCAL
	Water-powered toothbrush	Sterling Drug Company
	Water-powered appliances	" " "
	Expert witness GAF v/Airequipt	GAF
1978	Medical trailer	Amherst Corporation
	Can opener	Bonny Products
	Convertible bed mechanism	Castro Convertibles
	Ottoman mechanism	" "
	Convertible water bed	
	Expert witness GAF v. Airequipt	GAF
	Folding bed mechanism	Murphy Door Bed Corporation
	Prototype water-powered toothbrush	Sterling Drug
	Expert witness Famolare v. USM	United Shoe Machine, Inc.
1979	Build folding bed prototype	Murphy Door Bed Corporation
	Variable bicycle transmission	Scott/Mathauser
	Expert witness Famolare v. Thom McAnn	Thom McAnn
1980	Sailboat hardware	Navtec
	Long-range industry forecast	SOCAL
1981	Low emission range	BR Laboratory, Inc.
	Appliance power sources	Gillette

	RECON project	SOCAL
	Expert witness Enesco v. Himark	Himark
	Bicycle brake	Scott/Mathauser
	Microwave/infra-red oven	SOCAL
	Expert witness Willow & Reed v. Galleon	Willow & Reed
	Cruising trimaran design	Yachting Magazine
1982	Management consulting	BR Laboratory, Inc.
	Olympic bobsled suspension	Chelsea Design
	Solar collector array	SOCAL
	RECON program	"
	Fiber optics	"
	Room ventilator	"
1983	Filter research	BR Labs
	Expert witness Tyke v. Kolcraft	Kolcraft
	Expert witness Butler v. Block Drug	Block Drug
	Domestic service connections	SOCAL
1984	Residential gas piping	"
	Clean air range	"
	Expert witness Hennessey v. Ford Motor Company	Ford Motor Company
	Tape oral history of WDTA—Dave Crippen	Henry Ford Museum
1985	Expert witness Barsanti v. U.S. Tobacco	U.S. Tobacco
	Automatic sprinkler	Reliable Sprinkler
	Low-emission range	SOCAL
	India Mark II hand pump study	IDAC
	Minical rework	Alzeta Corporation
	Futures report	SOCAL
1986	Start autobiography	
	Research Futures Report	SOCAL
	Tape oral history of WDTA	Henry Ford Museum
1987	Dendrite snowmaking gun	Dendrite Associates
	Toilet-seat damper	Scariano
1988	Mark II snowmaking gun	Dendrite Associates
	Snowmaking nucleator	"
	Field trip Africa, India, London	World Bank
1989	Form STK Corporation	Dorwin Teague, Inc.
	Production MIII snowmaking gun	STK Corporation
	Stage II hand water pump design	Dorwin Teague, Inc.
1990	Instruction manual	STK Corporation
	Trouble shooting	" "
	Snowmaking sled design	" "
1991	Snowmaking tests, New England, Rockies, California, Australia, etc.	" "
1992	Mooring barge	Julius Petersen, Inc.
1993	Arrange to leave all papers to Henry Ford Museum	Henry Ford Museum

APPENDIX D

RESOURCEFUL DESIGN THINKING BEHIND A NEW CLEANER GROUP

BY WALTER DORWIN TEAGUE, Jr.

A collaborative Development in ENGINEERING FOR PERFORMANCE and DESIGN FOR APPEARANCE by Walter Dorwin Teague and W. D. Teague, Jr., Industrial Designers, with Chief Engineer Paul E. Frantz and Special Representative Robert Harroff of Apex Electrical Manufacturing Co., and also Division Manager for Major Appliances Howard Barber and Buyer for Major Appliances V. F. Peterson of Montgomery Ward.

REPRINTED FROM DECEMBER 1940 ISSUE OF ELECTRICAL MANUFACTURING

Resourceful Design Thinking Behind A New Cleaner Group

By W. D. TEAGUE, Jr.

Each of the four units in Montgomery-Ward's new line of vacuum cleaners is an engineering-design accomplishment in its own right. Together they form a well-integrated line with a family resemblance in both engineering for performance and design for appearance. A collaborative development by Walter Dorwin Teague and W. D. Teague, Jr., Industrial Designers, with Chief Engineer Paul E. Frantz and Special Representative Robert Harroff of Apex Electrical Manufacturing Co., and also Division Manager for Major Appliances Howard Barber and Buyer for Major Appliances V. F. Peterson of Montgomery-Ward

DEVELOPMENT of the new line of vacuum cleaners for Montgomery Ward represented, from the industrial designer's viewpoint, virtually the ideal design problem. This was for two very good reasons. *First,* the client imposed no coercive limitations based upon any preconceived ideas of what a cleaner should be or upon previous manufacturing procedures. The former line of equipment, which had been unchanged for several years, had served its purpose. This was not to be a mere revision but a completely new approach. *Second,* the arrangement between Montgomery Ward, Apex Electrical Manufacturing Co. (who build this equipment for Ward) and the industrial designers was such that complete cooperation was possible in all phases of the design from its inception until it was ready for production. In all, four separate cleaner designs were involved, two floor type, one tank type, and a small hand cleaner. The success of the designs may be gaged by the fact that sales of the floor-type cleaners alone show an increase of 120 per cent over those in the same period of last year.

The floor-type cleaners probably represent the most radical departure from accepted practice. Here the first step in the design was a market survey during which data were secured on performance, operating characteristics, methods of construction and various other selling features of all competitive products. From consumer reactions it was possible to determine the relative importance of these various elements for consideration in the new design. As a result of this study a number of objectives were set up; increased cleaning efficiency, better appearance, low cost, reduced operating noise, convenience in operation, lighter weight and maximum accessibility for servicing.

To achieve several of these objectives an entirely new blower principle for this type of vacuum cleaner was employed. Instead of an ordinary impeller with the scroll housing heretofore used, a radial type fan with air discharge back and around the motor was decided upon. One reason for this selection was the excellent operating characteristics of the radial fan which make it particularly adaptable to vacuum cleaner work where motor speeds are in the neighborhood of 12,000 rpm. Quietness at high tip speeds is an important quality of this type of blower. In the scroll type housing efficiency resulting from a sharp cut-off point close to the fan blades cannot be secured without excessive noise. The straight radial blower has no cut-off point to create noise and the possibility of large objects jamming between fan blades and cut-off point is eliminated.

Furthermore, this fan and motor combination adapts itself ideally to good dimensions for a vacuum cleaner. Height reduction is limited only by motor diameter

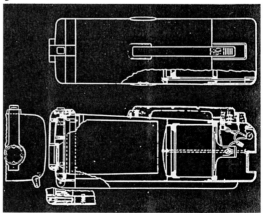

TANK model (A) offers pleasing appearance with good performance (see text). B—Details of the tank design.

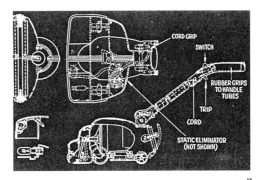

ENGINEERING the floor models called for some unusual features. A—These include a radial type fan instead of the conventional impeller type, greatly simplified construction and spectacularly good performance. B—Details of the overall engineering design. C—Motor specifications.

and width of air passage. The arrangement lends itself to notably simple construction from the manufacturing standpoint yet with good servicing accessibility.

As the drawings of the operating mechanism were made, the final form was developed progressively through plastic clay models, wood models, and finally operating models; thus mechanism and form, as evolved, are completely interdependent, and instead of pseudo-streamlining we have a design which is entirely determined by practical necessity. Actual streamlining of the air passage from the nozzle to the fan inlet, and around the motor to the outlet at the bag attachment, automatically results in a beautiful and functional shape. This shape is further accented by the lines of the rubber protector strips which follow its central axis from front to back and are also functionally essential. There is no need for extra housings to cover working parts or for added decoration to distract attention. Any such additions would detract from the symmetry of the working parts themselves.

Performance of the new model is all that can be desired. Extremely high vacuum for a floor cleaner is obtained. At the $1\frac{3}{8}$ in. orifice opening, which most closely approaches actual working conditions, it will develop a vacuum of $7\frac{1}{2}$ in. of water, an increase of 85 per cent over the previous model and 46 per cent higher than any other floor cleaner tested. At the same time operation is relatively quiet and the weight is $2\frac{1}{2}$ lb. less than the old model. The height of $6\frac{15}{16}$ in. to the top of the rubber protector strip is extremely low for a cleaner of this capacity.

TEST DATA—FLOOR CLEANER

	VACUUM IN IN. OF WATER			
	Sealed	$\frac{1}{2}$ In. Orifice	1 In. Orifice	$1\frac{3}{8}$ In. Orifice
New model (high speed)	15.31	13.31	10.00	7.50
New model (low speed)	8.37	7.06	5.06	3.62
Old model (high speed)	8.87	8.06	5.87	4.06
Old model (low speed)	4.12	3.87	2.67	1.87

Once the basic principle and shape had been decided upon, the next step was to determine the most efficient partition of the structure for accessibility and simplicity of manufacture. The cleaner body consists mainly of two large die castings joined on a horizontal line in the plane of the motor axis. A third small die casting forms the lamp cover and inlet to the impeller. A high degree of accessibility results from this arrangement. Removal of the small front casting which snaps in and out of place gives access to the lamp bulb and belt which drives the rotating brush. Steel springs in the small casting contact lugs on the two main castings and hold it in place. Felt and rubber sealing strips are used at this joint. Four screws and two dowel pins join the two large castings and removal of these four screws allows access to motor, fan and entire air passage. No gasket is necessary at this point.

An inherent virtue of this method of partition is the fact that the three castings become straight-draw pieces, without undercuts or other molding complications. The upper main casting forms a strong arch support for resistance to impact and for the handle and bag attachment. To the lower casting are assembled motor, brush, light, wheels, static eliminator and electrical connections. A rubber bumper strip is fastened to front and sides of the nozzle and additional protection is supplied by the beaded rubber strip running from the lamp aperture back over the upper main casting along the center line. This strip is in two pieces and contains a molded-in metal insert, supplied with hollow rivets at intervals for attachment.

The specially designed motor-impeller unit is another example of pleasing appearance resulting from the right adaptation of the design to its function. Since it is, in fact, part of the air passage, the exterior shape of the motor has been made to blend with the impeller and the surrounding air flow. Two die castings form the motor housing and a phenolic plastic cap at the rear, held in place by a single screw, allows access to the brushes. A tongue-and-groove type construction is employed between the plastic cap and the die-casting to seal this joint effectively. Two sealed ball bearings support the armature shaft, the front bearing being considerably heavier to take care of the impeller and overhung belt drive while a special loading spring behind the rear bearing takes up end play. The dynamically-balanced impeller is also die cast and employs tapering, backward-curved blades in accordance with best practice for radial fans.

Inlet and outlet for motor-cooling air are incorporated in the single rubber-mounted motor support.

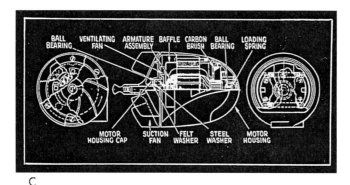

C

The motor has its own cooling fan at the forward end, separated from the rest of the motor by a steel baffle. The cooling-air inlet is prevented from coming into too close contact with the floor by a stamped plate on the bottom of the cleaner. First passing over the flanges of the plate, the cooling air enters the motor through the rear of the motor support, passes between the armature and field coils, through the center hole of the baffle and into the cooling fan, whence it is directed outward through the front aperture of the motor support. It will be noted that the motor support is tapered to the rear and set at an angle to conform to the helical path of the moving air. The motor wiring is taken in through the rear or inlet aperture of the support. Another feature which permits the motor to take a correct aero-dynamic shape is the special spring arrangement for the brushes. As indicated in the sectional drawing of the motor, the brushes are set in brass guides fastened to laminated plastic bridges on the main motor housing. Instead of having the usual coil compression springs, the brushes are connected by means of two tension springs, acting on fibre yokes, each of which has a tab set into a depression in the top of the brush. Thus the desirable small diameter is secured at this point without the usual projections.

The stamped plate on the bottom of the cleaner, held on with self-tapping screws, also houses the electrical connections. A removable laminated plastic junction plate set into cast-in slides receives the wiring from motor, 2-speed switch, static eliminator, light bulb, and handle. The light bulb in this unit is placed in a compartment just forward of the revolving brush. Bulb socket is molded plastic set in a soft rubber piece which also forms the side bulkhead of the compartment. This effectively seals the bulb compartment from the air stream around the brush.

The "Supreme" or best model of the line employs a two-speed motor of the tapped field type, a lower speed being useful for small rugs and places where too much suction is a detriment to easy manipulation. The speed selector button switch is mounted on the lower main casting for foot operation while the on-and-off switch is placed on the handle. This switch is operated by sliding the thumb backward or forward over the top of the handle giving particularly easy operation.

Also on the handle is a finger trigger for handle positioning, and this is connected with a lever working on a three-position ratchet fastened to the upper main casting. A spring at the handle joint acts to counterbalance the weight of handle and bag. Another concealed spring in the bag itself ties the top of the bag to the handle by means of a flexible cord connection so that the bag is always flat against the handle. An unusually large opening (3⅞ in. diam.) is supplied in the body of the cleaner for the bag connection. Reduction of velocity at this point increases efficiency and the wide bag mouth makes emptying easier.

The usual attachments are available for both floor cleaners. These are connected to the nozzle by means of an adapter which snaps over the bottom of the nozzle plate. This obviates the necessity for removing the belt from the driving pulley in using the attachments.

As previously mentioned, there is a companion floor cleaner known as the "DeLuxe" which is offered at a lower price level. This machine is quite similar in essentials but various features have been eliminated or simplified in order to further cut manufacturing costs. While the same basic motor and fan are used, the two-speed feature has been eliminated. Depth of the nozzle skirt is cut down and handle tube is painted instead of being chrome plated. Instead of the finger trigger for handle positioning, a foot lever is used which eliminates several parts from the handle. An exterior bag spring replaces the concealed spring and cord arrangement used on the "Supreme" model. The color of the "DeLuxe" is brown while the "Supreme" is gray-blue.

ANOTHER DEPARTURE FROM TRADITION

SEVERAL new features are incorporated in the tank type cleaner. A section approaching a square shape is employed for the body, which permits the use of low die-cast runners fastened directly to the body shell in place of the wire runners or wheels usually found on this type of cleaner. The shell is a single wrap-around steel sheet lap-welded.

The rear end bell, front end ring and front end bell

INDUSTRIAL Designers Walter Dorwin Teague and W. D. Teague, Jr., match a model against drawings.

are aluminum die castings with a bright finish. The front end bell is hinged to the end ring on extensions of the runners. A push button at the top releases the end bell, and undercuts on the vertical rib aid in opening for bag removal.

The carrying handle on the top of the unit is combined with a switch of the push-on, push-off type. Thus it is only necessary to step on the handle to start or stop the cleaner. As shown in an accompanying illustration, the handle is pivoted in front with a coil spring acting to raise it at the rear. Action of this spring is limited by a steel yoke. In front of the handle pivot a rubber boss is fixed so as to force the handle down and prevent rattles. When lifting the cleaner, the handle opens a little farther against the action of the rubber boss, thus allowing more finger room.

A $\frac{1}{4}$ hp. motor drives the dynamically-balanced fan. As is the usual practice in tank-type cleaners, a two-stage fan is employed. Due to efficient design and the extremely powerful motor, this cleaner also develops a higher vacuum than any other tank cleaner tested. The motor and fans are built in a single removable cylindrical-shaped unit. This is located and supported by a thick rubber ring at either end which provides the necessary seal and at the same time provides a resilient circular support at each end to insulate the moving part from the body of the unit.

Two long studs in the rear end bell run forward to a bulkhead near the center of the cleaner. These hold the rear end bell in place and the motor fan unit is supported between the rear end bell and bulkhead. The bulkhead is also formed-in to hold the circular filter.

The motor of this unit is of very open construction and since the primary air is cleaned by first passing through the dirt bag and then through the filter, it is used for motor cooling. Sealed ball bearings are provided for the armature shaft. The two impellers are fabricated of light-gauge aluminum, the backward-curved blades having tabs which fasten them to their discs. These units are dynamically balanced.

Wiring from the motor is shown in the sectional view of this unit. As on the floor cleaner, a static eliminator is built in. A rubber molded connector on the cleaner end of the 20 ft. cord allows it to be removed.

An ingenious method of hose fastening facilitates this operation. Thus, the hose coupling consists of two die castings fitted over a steel tube which in turn is cemented in place in the flexible hose. The fit used allows the hose to be swiveled freely at all times in the coupling. The knurled die casting is threaded on the inner member. Also fitted on the inner casting are a rubber washer, backed by a loose steel washer on either side, and a coil spring. On the end of the inner casting is a projecting steel pin. Four lugs are cast in the socket in the end bell which contact this pin. Upon insertion of the hose coupling in the socket, the knurled casting is rotated clockwise until the pin contacts the nearest lug. At this point the inner casting remains stationary and further rotation of the knurled section causes it to advance on the threads, compressing the rubber washer so that it expands laterally and causing the spring to ride up on the inclined shoulder. The spring expanding into the undercut in the socket prevents the coupling from pulling out while the expanding rubber washer makes a good air seal. Never is more than a half turn necessary to complete the coupling operation.

The finish on the body of the cleaner is in gray-blue fabric similar to that often used on luggage. This material has an excellent appearance and will stand a great deal of wear and tear. A matching blue color is used on the nameplate background and on the various accessories supplied with this cleaner.

TO MEET COST COMPETITION

IN THE small hand cleaner it was decided to use the same fan and motor principle as on the large floor cleaner. Two simple straight-draw die castings joined along the longitudinal center line of the unit form the nozzle, main body, bag attachment ring, handle and switch housing. This makes an extremely simple and practical arrangement for this type of equipment which is highly competitive and requires very low manufacturing costs.

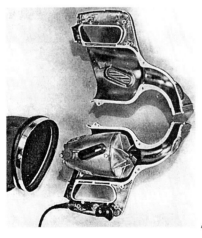

SPLIT vertically (A) the hand model (B) is a good lesson in competitive supremacy. Low cost is here essential to successful engineering-design but compromise need not be resorted to.

A   B

MONTGOMERY-Ward's DeLuxe and Supreme models differ more in performance and price than in general lines although different color schemes are used. Supreme (left) has a two-speed motor and has extra appearance touches. DeLuxe has a one-speed motor and is somewhat plainer.

TEST DATA—TANK CLEANER

MODEL	VACUUM IN IN. OF WATER				
	Sealed	½ In. Orifice	1 In. Orifice	1⅜ In. Orifice	Wide Open
1940	44.5	30.0	7.5	2.5	0.3
1930	16.8	10.0	4.0	1.8	Neglig.

As in the large floor cleaner, the motor has its own ventilating fan. Inlet and outlet are placed on opposite sides in the form of hollow bosses cast integrally with the motor housing and placed on a slant to conform with the helical air stream. These hollow bosses also act as the motor supports, rubber gaskets being placed between them and the main die casting. Small grilles cast into the sides of the unit improve appearance and protect the motor.

Entering through the air inlet the air passes to the rear of the motor and then forward between the field coils and armature. It then passes through the central hole of a baffle and to the steel motor impeller. From the impeller the air is forced past the cut-away side of the baffle and out through the outlet boss.

Prelubricated bronze bearings are held in place by pressed steel caps riveted to the motor housing diecastings. These bearings have a spherical outside contour which makes them self-aligning, thus facilitating general motor manufacture. As in the large floor cleaner, thought has been given to the contour of the motor housing to make it conform with the primary air path. In this case exterior brush bosses are employed but these are given a stream-lined shape and set at an angle to the motor axis.

In addition to the motor, the nameplate, switch, strain relief, and all wiring are located and held in place between the two main castings so that no other fastening means are necessary. Eight screws located at strategic points around the joint hold the two main castings together, no gasket being required.

Full advantage of the height of the unit is taken for the bag opening which measures 4⅜ by 6 in. This large opening is a feature which eliminates the necessity of a second opening in the bag for cleaning. Two polished aluminum die-cast clamps at the sides, working on pins in the bag ring, hold the bag in place.

Total weight of the hand cleaner is 5½ lb. Of this weight 4½ lb. are in the motor, fan unit, switch and wiring. The two main castings, bag assembly and bag clamps weigh approximately a pound only.

HERE IS THE POWER UNIT

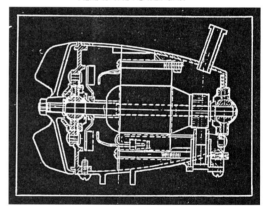

Appendix E

Dorwin Teague's Report to Edsel Ford Recommending a Ford Sports Car
13 November 1940

POSSIBILITIES OF FORD "SPORTS" CAR DESIGN

"A FORD CONTINENTAL"

For many years the American motor car industry has not offered anything in a small car other than the strictly stock low-priced models. Of late the convertible models, which have now almost entirely replaced the strictly open models, have advanced in price over the closed models and now run considerably higher ($900 to $1,000 as compared to the base prices of $600 to $700). These however have the same chassis and motors as the closed models. One Plymouth model this year is selling for $1,400 and offers special windshield and body trim but still no definite improvement in appearance or performance.

The European market, on the other hand, has always produced a number of smaller cars of very high quality as to appearance and performance. Prices vary from medium to very high but all endeavor to offer something better than the "family type" in the way of appearance, performance and handling characteristics. Many consider this tendency to be a result of the high horse power tax imposed abroad, and this undoubtedly has some effect. However since most of the true "sports-car" types are sold at prices over $2,000 the cost of the horse power tax is not a primary consideration to these owners.

The primary reason for this market is that the European buyer, as opposed to the average American citizen, feels that best handling, performance, and appearance characteristics may be obtained in an automobile of small or medium size. In many of these models certain comfort features are sacrificed. The European sports car driver does not care whether his car can carry three or four in the front seat or not, and does not object to more difficult entry and exit. He feels that a form-fitting bucket type seat is more comfortable on a long run than a wide flat seat.

On the other hand he places much more emphasis on ability to corner well; good driving visibility (on almost all true "sport" models both front fenders may be seen from the driving position); easy, direct steering for quick manipulation in emergencies; low center of gravity and similar characteristics. In appearance he prefers a lower and lighter effect rather than massiveness. Springing must not be soft to the point of producing roll and sway upon a deviation from a straight line, and braking must be very good. All these characteristics may be summed up by stating that the "sports" car is designed to travel faster and more safely on any type of road than the "family" type of automobile.

Examples of this type of car are the Alvis, Acedes, Bentley, Jaguar, Lagonda of England; Bugatti and Delahaye of France; and B.M.W. and Mercedes of Germany.

As stated above, some of these models are lacking in some of the comfort characteristics of even the cheaper American cars. No true "sports" model can carry as many people comfortably as a popular-priced American sedan. Some, not all, are less comfortable at low speeds on a bumpy road. On all these upkeep is more difficult and expensive for the non-mechanical minded due to inferior mass production technique. For the same reason reliability is not so great in most European types.

Any one who doubts that the foreign "sports" car excels in the aforementioned desirable characteristics has only to refer to records of recent races in this country or abroad in which American and foreign stock types have competed. This is especially true of the road race which more closely approximates normal driving conditions. The special racing car itself, either American or foreign, is proof of the fact that speed, safety, and desirable handling qualities require a moderately sized unit. This, of course, has nothing to do with the straight-away record cars, which are designed only for traveling at speed in a straight line for a few miles on a specially prepared course.

It is not possible, of course, that this type, as now built in Europe, will ever have any wide appeal in this country. However, the existence of a market for a modified version of this idea seems definite. This modification would take the form of a special model of a low priced car. Several reasons point to the existence of this market: First, the fact that a large number of convertible models at quite an advance in price, are sold. Foreign cars, even at very high prices, with repair facilities limited have a small but steady market. Younger people, college students and others who are not primarily interested in maximum comfort to the exclusion of certain other features almost invariably react favorably to this type of car; although at present market conditions prevent most of them from realizing any desire to own one.

The Plymouth special phaeton mentioned above is a first step toward a car of this type. Of the three popular low priced cars it would seem that Ford is best qualified to bring out a special model. For some time the Ford has enjoyed the reputation of giving the best performance of the three leaders. Fords are usually the choice of the young open-car buyers, many of whom could easily afford a more expensive car. This is probably because it comes closer to the true "sports" model than any other.

This car would not entail a great amount of chassis changes, motor changes, or even new body parts. The Ford chassis might be retained as is except for possibly stiffening spring action and limiting its travel. Larger dash-controlled shock absorbers would give a more solid and safer riding action for faster driving and cornering. Stiffer, shorter spring action would also help this factor in addition to lowering height of body. Another worth-while feature would be the use of smaller diameter wheels to further decrease height as on the special Buick at the New York Show. A more direct steering ratio could be fitted to improve handling. On this job accurate and quick steering would be more desirable than extreme ease in parking. At the same time the steering post could be given an increased rake and with this and a slight modification in control pedals the driver's seat could be moved further back and lowered. Another valuable chassis modification would be the reduction of unsprung weight by tubular front axles and lighter rear-end parts.

Motor modifications would not need to be extensive. Perhaps the option of the Mercury motor would be a good feature. Optional higher compression heads or special four-carburetor manifolding might also be offered. Another feature which would be very valuable would be a slightly better motor and motor compartment finish than on the standard line. This would not cost much but would mean a lot to the type of purchaser who would buy this car.

Body changes would entail the most new parts but do not mean entirely new dies by any means. Probably the four passenger convertible body type is the best to consider. By employing sufficient ingenuity many body parts might be adapted with slight modifications. For instance, Zephyr fenders might be employed but would be set much lower than on the Zephyr. Certain parts would have to have special dies but these would be kept to a minimum. An extremely low, well-proportioned effect similar to the Continental Zephyr on a smaller scale, would be sought after. If done well enough this style body might be retained for several years without major changes.

                                            W. D. Teague Jr.

WDT Jr:RM

November 13, 1940

CHASSIS MODIFICATIONS

5.50 - 16 tires

Add leaves to springs and flatten to limit action, front and rear

Add Andre' telecontrol shock absorbers (or equivalent)

Use tubular front axle

Cut weight of rear end if possible

Extend clutch and brake pedals to rear by amount necessary (2" ?)

Special low ratio steering gear box and Pittman arm. Increase rake of steering column, make adjustable, special steering wheel and gear lever.

MOTOR MODIFICATIONS

90 h.p. motor

Polished aluminum or copper alloy heads. 7.75-1 ratio

Polished aluminum intake manifold

Chrome-plated cap nuts for heads and intake manifold

Baked enamel finish on special exhaust manifolds. Twin Burgess type mufflers and tail pipe assembly

Chrome-plate certain motor parts - semi-show finish in motor compartment.

APPENDIX F

Saturday Review

May 23, 1964   25¢

Design in America

ARTICLES BY
Katharine Kuh
Walter Dorwin Teague Jr.
Vincent Scully
Wolf Von Eckardt
Arthur Drexler

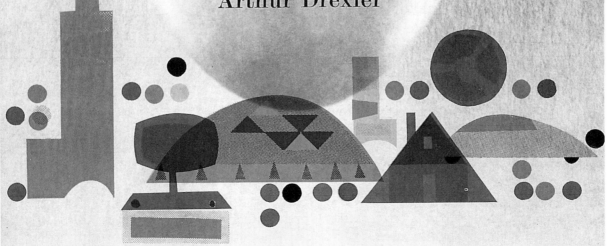

THE TWENTY BEST INDUSTRIAL DESIGNS SINCE WORLD WAR II

By WALTER DORWIN TEAGUE, Jr.

IN ANY ATTEMPT to pick "best designs," the biggest problem is to arrive at uniform criteria by which excellence can be measured. In most previous projects of this sort, the choice has been made by submitting ballots to any designers the organizers could talk into participating. In the present "contest," however, SR's editors thought it might be interesting to see what designs one man would select. The only ground rules—and these were established solely to limit the field—were 1) that the products should have been designed since World War II and 2) that they should not exceed twenty-five. (The final number turned out to be twenty.)

In some ways it is easier to justify a choice when it is a personal one. After all, if I am asked why I've included a particular item, I can always say simply, "Well, I just like it." On the other hand, I found that picking twenty-five truly classic items designed since World War II was quite a challenge. I went along at a good clip until I got to about seventeen, and after that it got slower and stickier. There were several designs (I've maintained the order in which they occurred to me) that I put down without question. I remember, for instance, the first time I saw the Honda Model X sitting at a curb (I think it was in 1960 in Bermuda) and I knew immediately that there was a classic design and that it should be part of anybody's list. But after reaching twenty, I discovered that I was dragging items in with increasing reluctance. It seemed reasonable, therefore, to stop there.

Most designers, I suspect, keep a conscious or subconscious list of this kind. In any event, it is obviously a personal choice and, I hasten to add, not necessarily the choice of Walter Dorwin Teague Associates, or indeed of anyone else but me. In fact, I'm sure that some of my partners will disagree with some selections as violently as my competitors will.

An obvious question will be: "Why such a large percentage of sports items?" Fair question. This is partly personal preference, but mainly the result of the fact that more design talent is applied to sports articles than to anything else. Particularly in the United States, our leisure-time equipment is much more important to us than the everyday furniture of our lives. (This is why, perhaps, some business executives will spend considerably more time, thought, and effort in improving their golf games than in improving the techniques of their businesses.)

In any event, here are my choices:

1. Karmann Ghia Coupe: This is a pretty obvious candidate for anyone's list. (The VW beetle is out by the ground rules—designed before World War II.) The Ghia looks just as good today, nine years later, as when it first came out.

2. Head Skis: A fashionable saying in America is, "We're pretty good when it comes to cheap mass production but you have to go back to the old country to get real craftsmanship." "Heads" are designed in the U.S. and built in the U.S. and, in spite of U.S. labor costs, high prices, and high duties, sell like hotcakes in Switzerland, Austria, and elsewhere.

3. Scott Ski Poles: Ask any good skier, whether he is French, German, Austrian, Swiss, or American, what he considers the best ski poles. He'll say "Scotty's" every time. To a non-skier this is hard to understand. They say, "What can you design in a ski pole? It's just a shaft, a strap, and a basket. Besides, those Scott things look sort of clumsy anyway." The racing skier says, "O.K., pick it up and wiggle it back and forth. [It's like wiggling a light straw.] Now look at the grip; it's got finger grips, a heavy pommel to improve the grip and lessen injuries, and it's canted forward so that when I make a slalom turn I don't have to cock my wrist as far. It's stronger than other poles and the basket is more rugged and easily replaceable. Notice the strap; it's adjustable so that I can get it to give just the right amount of support." In the five years since Scott decided to go into the pole business in Sun Valley, he has become the most copied pole manufacturer purely by fine, intelligent design and good, honest workmanship.

4. Triton Sloop: When I first walked into the 1959 boat show I had no idea of buying a new boat. Among the sailboats was a twenty-eight-foot fiberglass sloop called the Triton, made by an outfit called Pearson that no one I knew had ever heard of. I went back three times during the show, and on the fourth visit I wrote out a check for a down payment. I got Triton No. 8. Since then they have built more than 500 Tritons, a fabulous number for a cruising boat. Carl Alberg of Marblehead hit just the right combination of minimum boat with comfortable accommodations, a

1.

fast, delightful sailer with plenty of living room that still manages to look sleek and beautiful, in the water or out.

5. Bohn Calculator: A typical "product design" subject for an industrial designer. You couldn't redesign it for the next year's market without spoiling it.

6. Scott Paper Towel Rack: Another one that looked exactly right the day it came out. Not a useless line or contour anywhere. A real classic.

7. Honda Motorcycle: Every detail was studied to make it harmonious. You have to see it to appreciate its excellence.

8. Russian Single Shot Target Pistol; 9. Russian Autoloading Target Pistol: I first saw these in the Russian Pavilion in Vienna in 1957. They're both so terrific I put them in together. Note the peculiar low barrel on the automatic; there is no recoil lifting during rapid fire.

10. Questar Telescope: Another answer to the person who says we can't produce fine craftsmanship in the U.S.A. Designed and built in the little Pennsylvania town of New Hope, this beautiful little telescope is the finest instrument of its kind in the world.

11. Boston Whaler Outboard: All honest designers and real boat lovers have reveled in the meteoric success of the Boston Whaler, brought out in 1958 when, if you didn't have tailfins on your boat, you were passé. It has steadily forged ahead in sales and is now the most widely copied boat in the world. It's beautifully and honestly constructed and, although not cheap, it's worth the price.

12. Boeing 707: Here is American know-how, design, and craftsmanship in an unmistakable combination.

13. Eames Chair: This design set a trend that is still on its way. It seems like a fairly obvious design, but it had to wait until the materials were ready for it.

14. Olivetti Typewriter: Another example of how the Italian designers took off after the war.

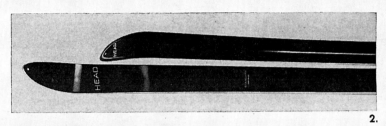
2.

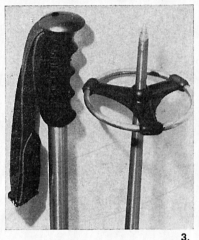
3.

4.

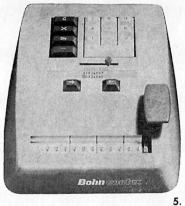
5.

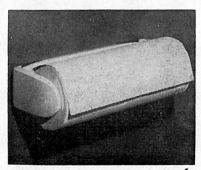
6.

7.

SR/May 23, 1964

15. Ponti Toilet: Those Italians again! If anyone has said thirty years ago that a toilet could be beautiful, he would have been laughed at. Well, I think it is.

16. Singer Vacuum Cleaner: Good design, well thought out from the "human engineering" standpoint. A fresh approach to the solution of a well-worked-over problem.

17. Porsche 904: Porsche cars are ugly in the opinion of many. But Porsche has never made any concession to current trends, and the factory still faithfully carries out the philosophy of "Let the form follow the function." This competition coupe is forty-two inches high, weighs about 1,400 pounds, and delivers about 180 horsepower with its four-cylinder air-cooled engine.

18. Wegner Chair: The problem with Scandinavian furniture is that there are so many good ones from which to choose. This was the inspiration for a whole school of furniture, a once-in-a-lifetime design.

19. Ericophone: This is a basic and clever solution to an old design problem. It's the kind of product that will grow on the user as he appreciates its convenience.

20. Carlsberg Beer Bottle: With all the talent that goes into packaging design, you'd think it would be easy to pick a candidate here. Not so—there have been good designs but no one so outstanding as to deserve to join the ranks of the twenty best. Just in time, Carlsberg came along with this new bottle. I think time will justify this selection.

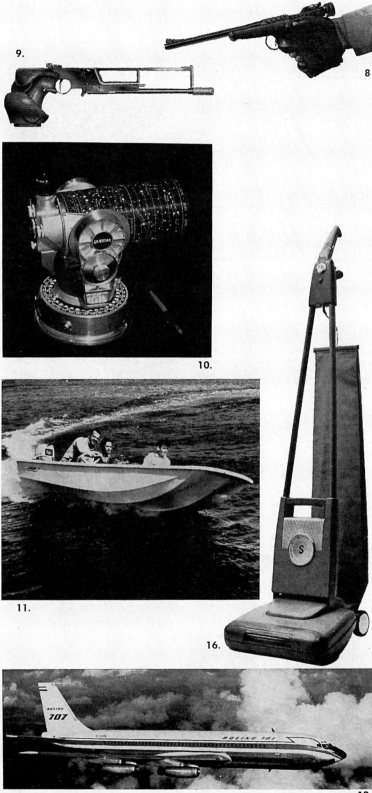

SR/May 23, 1964

18.

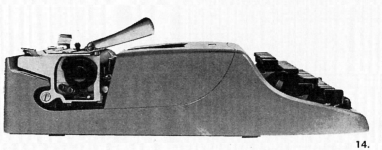

14.

17.

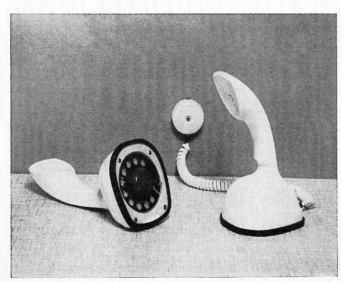

15.

19.   20.

Index

This book has been designed so that photographs appear on the same page as the relevant text. Therefore there is no separate illustrations index.

Abbott, Mary Lee, 88, 98
Ackerley, Bob, 107
Adams, Robert, 18
Addie (German Shepherd), 88–89
Advertising Club, 17
Airesearch, 159
Alberg, Carl, 150
Amateur Yacht Research Society, 113, 150, 157
American Horticultural Society, 140
American Institute of Aeronautics & Astronautics (AIAA), 162
American Power Boat Association, 107
American Saint-Gobain, 160
American Thermocatalytic, 163–64
American Underslung, 34
Ames, Allen, 154, 183
Amilcar, 46, 47
Anderson, Howard, 90
Animal Crackers, 22
Andrews, Bob, 217
Andrews, Dave, 178, 180, 206
Apex Electric, 69
Atoms for Peace Exhibit, 134–37
Atterbury, Grosvenor, 10–11, 17
Austin Healey, 130, 131
Auburn Automobile Company, 24; Convertible Sedan (1929), 44
Auto Union, 57–58
Automobile Racing Club of America (ARCA), 45, 59, 60, 64

Baby Brownie, 22
Baker, Talbot, 153–54
Balcom, Ronnie, 20
Balfour, Nicholas, 75
Ball, Edward, 14
Bare Boat Chartering, 152
Barney, W.J., Corporation, 158–58
Bascom, Bill, 169
Baum, Joe, 159
Beard, James, 164
Bel Geddes, Norman, 62, 158
Bellows, George, 13
Bemelman, Ludwig, 116
Bentley, "Blower," 99
Bendix Aviation Corporation, 74–87, 90–96, 98, 100–07, 116–17; projects, 79–83, 85–86, 91–93, 95–96, 104, 106
Besler, Bill & George, 77
Bien, Ward, 88, 96
Bishop, Peter, 187
Black, Fred, 72
Blackbirds, 22
Blackhurst, Rod, 20
Blood, Bob, 125, 135, 140, 158
Bloemhard, Walter, 150–51, 153–54, 157, 169
BMW 328, 75
Bohlen, Chip, 122, 124
Boeing B17 Cabin Supercharger, 74, 76; Stratocruiser, 116
Bomba, 152–53
Boucher, 34
Braman, Grenville "Gee," 60–61
Braman-Johnson Flying Service, 104
Braymer, Larry, 165
Breech, Ernie, 90, 98
Brick House & Barn, 28, 30, 44
Bridgehampton Track (1958), 133

Briefcase Fire Escape, 202
Brophy, Jack, 14, 118, 173
Brown, Shellman, 154
Bryant Heater Company, 48, 52
Bush, Vanevar, 24
Butler, Scrib & Anita, 87
Bugatti, Carlo 46
Bugatti, Ettore, 46, 65, 99
Bugatti, Jean, 46
Bugatti Automobiles, 46, 47; Type 37, 132–34; Type 41, 47; Type 51, 64–65, Type 52, 47; at Shelsley Walsh, 105
Bugatti Owners Club, 132

Cadillac Model K (1907), 15
Calkins & Holden, 14
Calloway, Cab, 60
Calspan, 206
Campbell, Jim & Margot, 75
Campbell, Malcolm, 99
Caracciola, Rudi & Alice, 126
Carlton, Hank, 24, 26
Carter, Rick, 211
Castro, Bernie, 197–98
Castro Convertible Company, 197–98
Caterpillar Tractor, 52
Cathedral School of St. Paul's, 20
Catlin, Guy, 18
Cervit, 164
Chevrolet Roadster (1924), 20–21, 23; Special DeLuxe (1941), 72
Chrysler Sedan (1926), 15, 23, 28; Airflow (1934), 50
Civil War Centennial Information Center, 148, 149
Clark, Art, 158
Classic Car Club of America, 37
Cocoanuts, The, 22
Collier, Sam & Miles, 59
Columbia Gas, Inc., 165, 195; Advanced Dryer, 176; Infra-Red Burner, 173–75; Smooth Top Range, 163–64
Columbia Mills, Inc., 119, 120; folding door, 119
Conrad, Carl, 118, 122
Consolidated Edison, 52
Cooper, Fred & Peyton, 26
Corcoran, Tom, 206
Cord L-29, 24, 25
Cotten, Joe, 148
Cotton Club, 60
Cousins, Norman, 165
Cowan, Clyde, 134, 137
Cowper, Vernon, 124, 172
Cramp, Charles H. (freighter), 20
Crosley, Powel, 53
Crosley Corporation, 53
Crowe, Dick, 88
Cruising Club of America, 146, 154, 215
Cummins, "Wild" Bill, 58
Cunningham (automobile), 17
Cunningham, Briggs, 133
Cunningham, Sam, 187, 195
Curtis, Morris & Safford, 196
Curtiss Oriole, 20
Curtiss Wright Corporation, 98

Deland, Dave, 142, 143, 144
Del Guidice, Frank, 118, 173
De Monge, Louis Pierre, 189
Dendrite Associates (snow-making gun), 206–07, 209–11, 215
Despot, 125, 126
De Vito, Danny, 191
De Vries, Lensch, 153–54
Dick, A.B., Company, 48, 50, 52
Dictaphone Corporation, 196
Disney, Walt, 158

Doble Steamer, 77, 78
Dobrick, Herb, 100, 103, 111
Donohue, Mark, 133
Doremus, Widmer, 18, 26, 31, 88, 89
Douglas, Michael 191
Dow, Wells, 154,. 178, 183
Dresser Manufacturing Company (high speed radial gas engine), 48, 49
Dreyfuss, Henry, 101
Drinker, Harry, 98
Dubin, Fred, Associates, 159
Duesenberg Model J, 75
DuPont Corporation, 52

EAA Air Adventure Museum, 189
Earl, Dick, 20
Eastman Kodak, 21–22
Easton, Karl, 19–20, 65
Eclipse–Pioneer (Bendix), 74–87, 90–96, 100–07, 116–17
Electrolux Corporation, 120–21
Ellington, Duke, 26, 60,. 124, 125
Elliott, Collier, 18
Ensign, Bob, 71, 119

Fabian, Ed, 152
Falise, Greg, 196
Federal Building (1939 New York World's Fair), 52
Fehon, Abbie Ann, 12
Fehon, Amy, 12
Fehon, Maude, 12, 15
Field, Jack, 125, 135, 158
Fisher, Freddie, 60, 104, 105, 139
Flanagan, Bob, 77, 100, 101, 104
Floyd-Wells Company, 52
"Fomento" (Puerto Rico), 150–52
Ford, Edsel, 63, 72, 73
Ford Motor Company, 52–53, 55, 72–73; Broadway showroom, 53–55
Ford "Sports Car" (DT design), 72, 239–44; Two-Door Phaeton (1931), 44–45; Woody Wagon, 117
Foreign Agricultural Service, 140
Forest Hills, New York, 10, 11, 14–15, 22
Fournier, Herb, 90
Fromm, Charlie, 169
Fuller, Avard, 122
Fulton, Robert (paddle steamer), 12, 13

Gaffney, Frank, 93, 94, 101, 108, 110
Galerne, André, 169
Gardiner, Hank & Winnie, 124, 145, 180
Gardner, Captain Edward, 137
Gardner, George, 125, 135, 166
Gas, Inc. Pavilion, 158–162
Gibbons, Llew & Helen, 146, 154
Gilhooley, Ray, 46
Glen Goin, 74, 75, 87
Goddard, Robert Hutchings, 90–91
Gollin, Sy, 171, 173
Gooding, Jack, 18
Goudy, Fred, 17
Gould, Milton, 172
Graham, Will, 154
Green, Bill, 171
Gregg, Dave, 74, 81, 90
Gregorie, Bob, 72
Guggenheim, William, 74

Hackenberg, Emil, 84, 87
Haggerty, Lamont, 154
Halley, McLure, 64,65
Halliday, Ken, 93, 101
Halsted, John & Nancy, 75, 113
Hammerskjold, Dag, 137
Hampton, Ben, Advertising Agency, 14
Hanson, Dr. Peter G., 157

250